VOICES
of the
EARTH

VOICES of the EARTH

and the explorers and pioneers who inspire us to listen

JO WOOLF

Royal Scottish Geographical Society

First published in 2025 by

Madwolf
DESIGN

Design & typesetting by Madwolf Design
madwolf.co.uk

Printed in the EU by **BooksLine**
Printed on paper from well managed forests using 100% renewable energy

British Library Cataloguing in Publication Data
A catalogue record for this book is available from the British Library

ISBN 978-0-9556968-6-2

Published for
Royal Scottish Geographical Society
Lord John Murray House 15-19 North Port Perth PH1 5LU
www.rsgs.org

Royal Scottish Geographical Society

The RSGS is a charity registered in Scotland, No. SC015599

The open road still softly calls, like a nearly forgotten song of childhood.

Carl Sagan, *Pale Blue Dot*

CONTENTS

FOREWORD

> Amongst the many Sciences... *Geography* is worthy of all praise: because it both affordeth matter of much delight, and likewise is of much use for the life of man... What delight must it give us, when we learn by it the Situation of Countreys, Towns and Rivers, as well as if we had viewed them with our eyes!
>
> Sir Robert Sibbald, 1683

In the proposal for his *Atlas of Scotland*, the country's first Geographer Royal, Sir Robert Sibbald, celebrated the practical value of geography, but also the pleasure that came from it, allowing people to experience different places as if viewed themselves.

This was something keenly recognized by Agnes Livingstone Bruce and John George Bartholomew two hundred years later when they proposed a Scottish Geographical Society to advance the study of the subject in Scotland, and 'to impress the public with the necessity and inestimable value of a thorough knowledge of geography.' The establishment of the RSGS is the starting point for *Voices of the Earth*: a time when popular interest in geography was growing alongside the expansion of Britain's Imperial possessions. In his speech to mark the opening of the Society, Henry Morton Stanley celebrated its role:

> not as a science to be relegated to your sedentary hours, but as a science brimful of interest to you as living men with warm sympathetic hearts, desirous of increasing your fellowship with humanity in general; [...] as a science wherein there is nothing barren or sterile to any man with eyes to see and ears to hear.

The heroic tales of exploration that Stanley and others told of the farthest reaches of the planet drew in large audiences to the Society. This was a rare opportunity for people in Scotland to hear directly about the new lands being brought into the Empire, hear tales of exotic peoples and places, and, perhaps, imagine the adventures that they themselves might have one day.

While our understanding of the Imperial mission has now changed, and perhaps our celebration of certain figures from this period has shifted to one of ambivalence or even disapproval, the draw of good writing about the world has lost none of its power. *Voices of the Earth* covers the period from the establishment of the RSGS in 1884 to the present day, reflecting changes in understandings of geography, place and environment. Over this period, the aims of the RSGS itself have changed too, and it now plays a leading role in education and policy supporting sustainable futures.

In her own explorations into the archives of the RSGS, starting from the basecamp of the Visitors' Book, Jo Woolf has charted this shift, from the explorers supported by, and supporting the commercial and Imperial ambitions of the state, to those intent on discovering the science behind the world's most challenging environments, to those wishing to bring the world's focus onto urgent issues, or just to communicate the pleasure of being in wild places to a wider audience.

Some of these figures we will know – or feel we know – already as their adventures are well documented. But Jo has found others whose efforts have been less celebrated. Some, like Gertrude Bell, Myrtle Simpson, and the Scottish Women's Himalayan Expedition, had faced the extra challenge of wanting to do things that weren't considered suitable or even possible for women. And, contrary to the lone heroic explorer narrative, she shows how many adventurers are dependent on the support and generosity of many others. The vital assistance of indigenous peoples and their in-depth understanding of the challenging environments being navigated comes to the fore here, from John Rae's acknowledgement of the Inuit's superior knowledge of the ice to Elizabeth Maruma Mrema's insistence of the importance of traditional African knowledges to the ecological support of biodiversity.

Voices of the Earth is a celebration of people who are passionate about exploring and protecting our planet's natural environments. Jo Woolf has brought together the accounts of a wide range of people from more than 140 years of the RSGS, looking in depth at their achievements, while at the same time learning more about them as people. She asks why are they attracted to these often remote and inhospitable places, what challenges did they face there, and how did they overcome them? Throughout the chapters, she probes at what keeps drawing them back. As readers, we too, it seems, are unable to resist the call of these voices.

Professor Jo Sharp
Geographer Royal for Scotland
St Andrews

PREFACE

When I became Writer in Residence at RSGS in 2015, I'd already spent about six months happily probing around in the Society's archives and collections, and I knew I'd only begun to scratch the surface. The number of people involved in some way with RSGS is unfathomable and ever-expanding, and I'm still coming across new names and new expeditions that demand investigation.

Many explorers and public figures, both historical and contemporary, were featured in my first book for RSGS, *The Great Horizon* (2017). *Voices of the Earth* is intended partly as a follow-up, and I have deliberately focused on a fresh set of people - the only (hopefully forgivable) overlap is Sir Ernest Shackleton, whose story is brought forward into the present day. In *Voices of the Earth*, however, I have tried to look a little deeper, to get a bigger picture, with every person I've included; as well as asking how and when, I've asked why, and in so doing I hope that I've gained some insight into the events and decisions that shaped their lives.

One of the quotes that I keep on my desktop is by Freya Stark, who never let anything or anyone stand between her and her wandering spirit: 'I know in my heart of hearts that it is a most excellent reason to do things merely because one likes the doing of them.'[1] That applies to so many people featured in this book, and it also sums up my own experience in writing it.

Jo Woolf FRSGS
RSGS, Perth
February 2025

[1] Freya Stark, *The Valleys of the Assassins* (1934)

I feel most at peace and most truly myself when I am out there questing, moving over the earth's surface. Then, all the noise and distraction of the world fades and you can become truly human again.

Alice Morrison, *Walking with Nomads*

INTRODUCTION

One of the most precious treasures in the archives at the Royal Scottish Geographical Society is the original Visitors' Book. Half-bound in leather and weighing over three and a half kilos, its well-thumbed pages have recorded the daily comings and goings of RSGS members, staff and visitors ever since 4th December 1884, when the explorer Henry Morton Stanley signed his name with a flourish at the top of the first page.

In those early years, some names recur frequently: Agnes Livingstone Bruce, for instance, daughter of the missionary David Livingstone; and John George Bartholomew, from the eminent map-making family of Edinburgh. Together, these two had actually provided the spark that kindled RSGS into being.

But scattered in between the regular entries are some names that really stop you in your tracks, because they were written by people who had just returned from all the most distant and inhospitable (and, at that time, still relatively unknown) parts of the Earth. Take the Antarctic, for example: Carsten Borchgrevink, having survived a whole winter on the Antarctic continent (his was the first expedition ever to do so) added his name on 14th November 1900.

The signature of another Antarctic explorer, Sir Ernest Shackleton, first appears on 26th May 1910. Shackleton had come to share his story of the *Nimrod* expedition with audiences in Scotland, slipping lantern slides of icebergs and penguins into the limelight projector and joking about the discomforts, while concealing a sense of disappointment that he'd not quite reached the South Pole. (Shackleton was no stranger to RSGS: he'd been employed as Secretary six years previously, and he still has an almost tangible presence here.)

But fast forward to the 21st century, and a new version of the Visitors' Book has been signed by visitors such as Doug Allan, a film-maker who has spent many seasons in Antarctica, braving the advances of boisterous seals to obtain the most breathtaking footage of the Southern Ocean and its wildlife. It has also been signed by members of the Endurance22 Expedition, who, in March 2022, defied formidable odds to discover and photograph Shackleton's lost ship, *Endurance*, at a depth of nearly 10,000 feet in the Weddell Sea.

This continuing celebration of human endeavour is symbolic of something that lies at the very heart of RSGS. Promoting and expanding geographical knowledge has always been one of the Society's core aims, and an effective way of doing this is by listening to people who have made extraordinary journeys. Sometimes these are stories of discovery, and sometimes they are stories of survival; they reveal personal courage, optimism, resilience and commitment. It is with good reason that the RSGS's annual programme of guest speakers is called Inspiring People Talks.

These stories tell us something about ourselves, too: about the nature of communities all over the world, and the ways in which people of vastly different cultures can find mutual respect, understanding and kindness. Myrtle Simpson, for example, took her children with her on many expeditions in the Arctic, believing that they would overcome any language barrier and thrive in the company of other families (which, of course, they did). Rory Stewart, walking through war-torn Afghanistan, found that villagers living in an atmosphere of instability and fear still welcomed a traveller, in keeping with their long tradition of hospitality.

In so many instances you get the sense that, beneath all the superficial differences, there is a deeper commonality. People who live a way of life that has kept them in closer touch with the natural environment tend to tread lightly and respectfully, sustaining themselves without making too many demands on the Earth. John Rae, tramping in snowshoes across Arctic Canada, understood that the people of the Arctic could teach him all he needed to know. Benedict Allen, pushing through the tangled rainforests of Papua New Guinea, turned to indigenous people for guidance. Under the scorching sun, both Alice Morrison and Mark Evans learned that the pace of desert travel is governed entirely by the cooperation of camels.

That's not to say that our own society in the western world was always supportive of people who smashed the mould and disappeared off towards the horizon, particularly if they were women. Monica Jackson, who travelled with her friends Evelyn Camrass and Betty Stark to the Himalayas in 1955, was quite clear about the opinions held by their male counterparts, who thought that a climbing expedition organised by women was a recipe for disaster. In 1900, Gertrude Bell bought herself a man's saddle and wrote to assure her parents about the respectability of her new 'divided riding skirt' before galloping off into the deserts of the Middle East, where she was free to make her own decisions about what she did and where she went.

And what drew these people away from the comforts of civilisation in the first place? When Shackleton wrote of 'the lure of little voices' he was referring to the eponymous poem by Robert Service in which a former traveller yearns to go back to the lonely trail. Other explorers have found a similar expression: the title of Colonel John Blashford-Snell's book, *Something Lost Behind the Ranges*, borrows a line from Rudyard Kipling's poem, 'The Explorer', in which a settler is unable to resist the call of unknown lands.

It's interesting to consider how explorers and travellers are drawn to different environments, but ultimately discover a truth that sounds or feels very similar: something akin to peace, fulfilment, and a sense of purpose that might be closer to a sense of belonging. Leo Houlding, for example, speaks of the mountains as incredibly powerful places; a century or so earlier, this opinion would no doubt have been shared by Alexander Kellas, for whom the mountains were both a fascination and a source of peace. Steering her kayak down river rapids, Sal Montgomery finds absolute focus in the moment, surrounded only by natural sounds.

Many of the people who feature in this book are, in some way, helping to increase our appreciation of the Earth's natural environments, their inhabitants, and the threats that they face. They are raising their voices in support of action and attitudes that will bring us back into a better and more harmonious balance with the Earth. For this to happen, communication and cooperation are key. In the words of RSGS's Patron, His Majesty The King: 'Change will come by working together and making it easier to embrace decisions that will sustain our world, rather than carry on as though there are no limits – or as though our actions have no consequences.' [1]

In writing this book, I have been very fortunate in being able to collaborate with some extraordinary people in the telling of their story. This has given me a huge appreciation for their courage, honesty, dedication and talent. Without exception, they are all embracing something that they believe in with their entire heart and soul. They are following a path that isn't always comfortable (very often quite the opposite), but they know for sure that it's right for them. I think we can benefit a great deal from their experience.

[1] A Speech by His Majesty The King at the Opening of COP28, Dubai, UAE, 1st December 2023, via https://www.royal.uk

N

POLAR

But from first to last the history of polar exploration is a single mighty manifestation of the power of the Unknown over the mind of man.

Fridtjof Nansen, *In Northern Mists: Arctic Exploration in Early Times*

Can any motor-car joy-ride give half the excitement of an afternoon behind a team of dogs?

Isobel Wylie Hutchison, *On Greenland's Closed Shore*

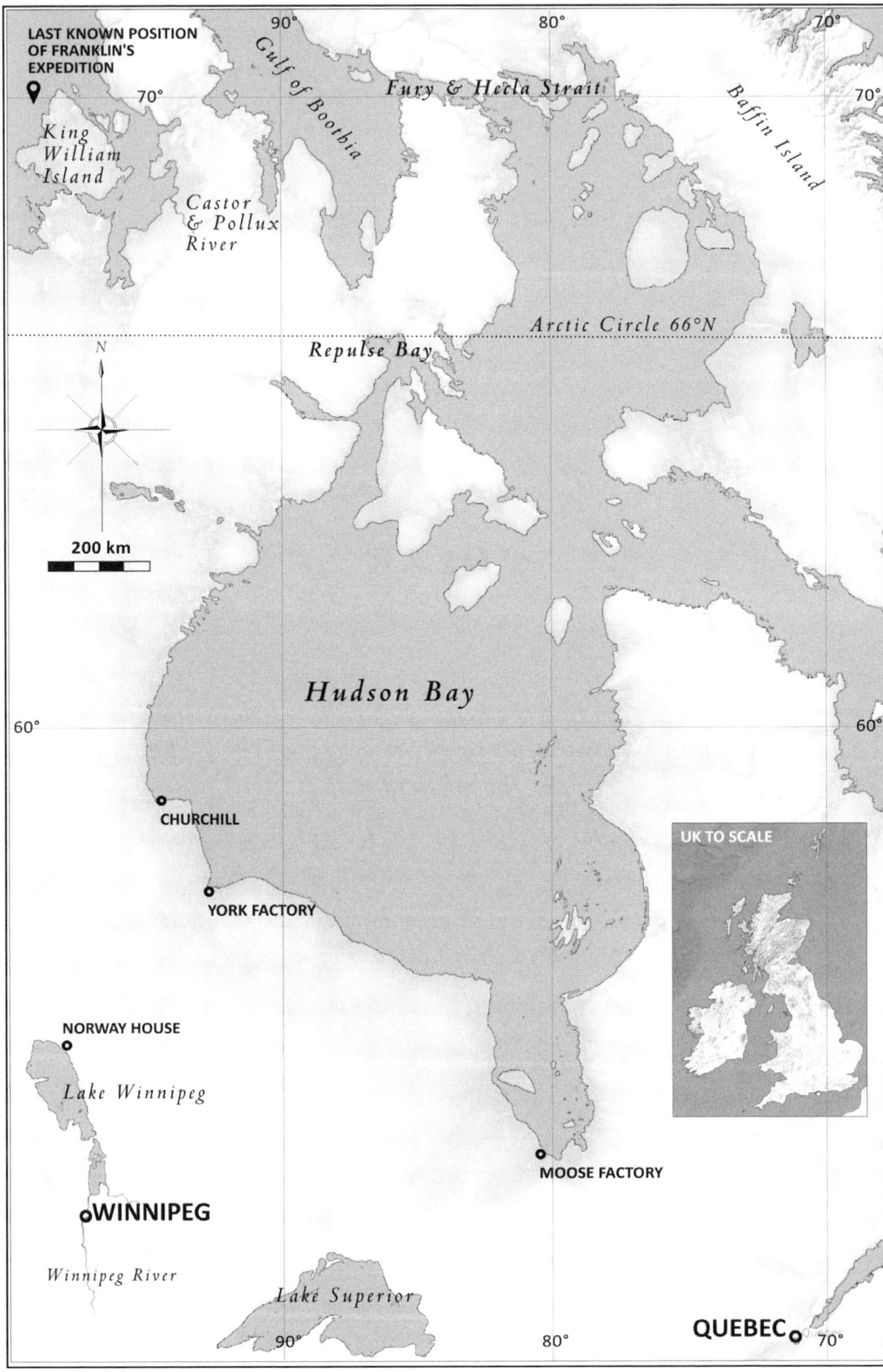

LAST KNOWN POSITION OF FRANKLIN'S EXPEDITION
King William Island
Castor & Pollux River
Gulf of Boothia
Fury & Hecla Strait
Baffin Island
Arctic Circle 66°N
Repulse Bay
N
200 km
Hudson Bay
CHURCHILL
YORK FACTORY
UK TO SCALE
NORWAY HOUSE
Lake Winnipeg
WINNIPEG
Winnipeg River
MOOSE FACTORY
Lake Superior
QUEBEC
90°
80°
70°
60°

John Rae

On 7th September 1845, a young clerk for the Hudson's Bay Company (HBC) was canoeing south-east on the Winnipeg River in the company of a group of fur-trappers and *voyageurs* who plied their trade over long distances in Rupert's Land, a vast territory later incorporated into Canada. Coming towards him, he noticed a lone canoeist who was heading north, a fact which astonished him because the season was late and 'all the canoes accustomed annually to go through these wilds had long since passed.' [1]

Out of mutual curiosity, both canoeists steered towards a convenient stretch of flat rock and got out to make each other's acquaintance. The HBC clerk introduced himself as Robert Ballantyne (he would later go on to pen a number of children's adventure stories under the name of R M Ballantyne). The north-bound canoeist revealed that his name was John Rae. Ballantyne knew him by reputation as a talented HBC doctor, a superb rifle shot, and one of the best snowshoe walkers in the region. In the flesh, Ballantyne observed that Rae was 'very muscular and active, full of animal spirits, and had a fine intellectual countenance.' [1]

After a few brief minutes the encounter was over, but the purpose of Rae's journey, which he had readily disclosed, preoccupied Ballantyne for some time to come. Rae was heading for York Factory, a lonely HBC settlement and trading post on the shore of Hudson Bay. There, he would equip an expedition to survey a stretch of Arctic coastline that still remained unmapped, in the hope of discovering a navigable channel between the Atlantic and Pacific Oceans. In short, Rae was going in search of the Northwest Passage.

He would equip an expedition to survey a stretch of Arctic coastline that still remained unmapped.

Twenty years previously, as a young lad growing up in his native Orkney, John Rae could only dream about such adventures. From his home at Hall of Clestrain, where his father was a factor for Lord Armadale, a local landowner, Rae tramped for miles across the hills and moors, shooting grouse and wildfowl for the family's larder, fishing in the lochs, and caring little about being outdoors all day in the foulest of weather. Hardiness and self-reliance ran in his blood.

Rae's father provided John and his three older brothers with an eighteen-foot sailing boat, and they became so skilled that they could race - and often beat - the pilot boats that went out to guide the ships arriving into Stromness harbour. For centuries, Stromness had been a last port of call for vessels embarking on long voyages: they would stock up on fresh vegetables and draw water from Login's Well in the main street.

In Rae's time, many of the ships coming into Stromness were supply vessels for the Hudson's Bay Company; they would spend a night or two in port before heading west across the Atlantic to Rupert's Land. This wilderness of lakes and rivers covered about 1.5 million square miles (3.88 million square km); when King Charles II granted it to the Hudson's Bay Company in 1670, it was named after his cousin, Prince Rupert of the Rhine.

With the departing HBC ships went hundreds of Orkney men, destined for a new life or at least a good few years of employment in an unfamiliar land. When John Rae was only six, his father took a new job as the HBC's agent in Orkney. His role was to recruit young men for service in Rupert's Land, and he could offer twice the wage that an agricultural labourer earned at home. It was a tempting prospect, and in the early 1800s, nearly three-quarters of the workforce of the HBC were Orcadians.

Not surprisingly, two of Rae's older brothers went out to Rupert's Land before him. Rae was impatient to follow them, but first, at the age of 16, he travelled south to study medicine and emerged three years later as a Licentiate of the Royal College of Surgeons of Edinburgh. In June 1833 he sailed out of Stromness as a ship's surgeon on board the HBC ship, *The Prince of Wales*; his intention was to return later that year, but fate, in the form of an impenetrable ice barrier in the Hudson Strait, intervened.

With the rest of the passengers and crew, Rae was forced to over-winter in makeshift quarters in the south of Hudson Bay. By the time his way home was clear the following summer, his skills as a doctor and outdoorsman had come to the notice of Sir George Simpson, Governor of the HBC, and he was persuaded to stay on at the HBC post of Moose Factory as a surgeon. Rae decided to embrace his new existence, which involved learning the skills of First Nation, Inuit and Métis people. Unusually for that time, he treated them as equals, and he chose to travel with them whenever possible. He found that their clothing, made from furs and other natural materials, was much better suited to the climate than his own. Immersing himself in their ways of hunting, fishing, sledge-hauling and snowshoe-walking, he gained knowledge and techniques that helped him to flourish in a notoriously hostile environment.

While the HBC owed its existence largely to the fur trade, written into its charter was an undertaking to seek the Northwest Passage. In Britain, the quest for this near-fabled waterway across the top of North America, with its promise of lucrative new trading routes, had become something of an obsession. In July 1845, Simpson decided, not without justification, that Rae was the ideal man for the job. First, he asked Rae to train as a surveyor and sent him on an overland trek to Toronto, over a thousand miles away, in order to find a tutor. Once he was suitably qualified, Rae's mission was to sail north up Hudson Bay and chart the unmapped section of the Arctic coast between Fury and Hecla Strait and the mouth of the Castor and Pollux River. He would take a party of 10 men, two small boats, and only a limited supply of provisions for the Arctic winter; when these ran out, they must live off the land.

Despite Rae's formidable reputation as an outdoorsman, he still had trouble hiring men to join him. He had already secured his trusted friend John Corrigal, himself an Orkney man, and Richard Turner, a Métis carpenter, but at Norway House on Lake Winnipeg he found that 'a report had got abroad… that the whole party, if not starved for want of food, would run the risk of being frozen to death for want of fuel.'[2] He persuaded five more men to take the risk, and in early October they arrived at York Factory on Hudson Bay in a freezing gale. Ice had already begun to form on the rivers, a sign that winter was closing its grip. Rae wisely opted to bide his time - and gather his remaining men - for a start the following spring.

Four months earlier, on 3rd June 1845, two Royal Navy ships had sailed out of Stromness harbour and headed across the Atlantic. Like Rae, they were engaged in a search for the Northwest Passage; but this was a maritime expedition on a grand scale, lavishly equipped and carrying with it all the expectations of the British Admiralty. There was no question of failure. The ships' officers counted themselves fortunate to be recruited, and the prevailing belief was that they would make it through to the Pacific the following year. Before their departure, the expedition's commander, the affable Sir John Franklin, went ashore to dine with Rae's sister, Marion, and her husband, Dr John Hamilton, at their house in Stromness. In mid-July, having offloaded surplus supplies onto an escort ship that had gone with them as far as Greenland, HMS *Erebus* and HMS *Terror* turned west towards Baffin Bay. A shouted conversation with the captains of two whaling vessels was their last living contact with the outside world.

The two wooden sailboats into which John Rae and his companions packed stoves, kettles, lamps, fishing nets and a small inflatable boat were about 22 feet long and sat low in the water. Rae had named them *North Pole* and *Magnet.* In June 1846 they made slow progress up the shore of Hudson Bay, sheltering from gales in the numerous small creeks, and hunting deer and wildfowl as they went. It would be a long, arduous journey, but they were in good spirits. As he lay awake in his makeshift bed on sacks of flour, listening to the occasional blowing of whales and the eerie calls of divers, Rae pondered his chances of success. He wrote: 'Little was known of the resources of that part of the country to which we were bound... Yet the novelty of our route, and of our intended mode of operations, had a strong charm for me, and gave me an excitement which I could not otherwise have felt.' [2]

Calling at the HBC's settlement at Churchill, where Rae recruited an Inuit interpreter named Ouligbuck and his son, William, they sailed right through the near-perpetual daylight when winds were favourable, landing on small islands to find fresh water, gathering seaweed and low-growing willows for firewood and catching salmon in their nets. Rae made careful notes about plant and animal species he encountered, and on a cloudless day he marvelled at a magnificent sun halo 'of the most brilliant colours, with four parhelia that rivalled the sun himself.' [2] The discomforts inflicted by wilder weather barely registered in his mind. On finding that his blankets had been soaked by a breaking wave, Rae merely observed that 'this, however, made them keep out the wind better, and did not... affect my rest.' [2]

By the end of July they had reached Repulse Bay* in the far north-west of Hudson Bay. A small group of Inuit men greeted them on the shore. Communicating through Ouligbuck, Rae learned that there was sea no more than about 40 miles away to the north-west, across a strip of land that consisted largely of rivers and lakes. This sea, although unknown to Rae at the time, was the south end of the Gulf of Boothia. He decided to transport one of his boats across the isthmus by rowing, poling and dragging, and persuaded some of his new Inuit friends to lend a hand.

Five days of strenuous exertion gained them the shore of the sea, and Rae launched his boat on it, sending a few of his men back to Repulse Bay overland. It was still August, but within a few days the sea was freezing around him, signalling the onset of another winter. He gave up any hope of pushing further north that year and returned to his companions. As the hours of daylight dwindled and snowstorms arrived, he set about laying in a store of fuel and food - fish, deer, hares and partridges - and studied the snow-houses built by the parties of

Inuit who visited his camp regularly to barter; they provided much better insulation than huts made of stone. Rae and his men built themselves some snow shelters, cleverly linked by internal passages, and prepared to sit out the long winter.

Christmas and New Year's Day were celebrated with a game of football, followed by a dinner of venison, plum pudding and a tot of brandy. Rae's men were in good health and spirits. He reflected: 'I do not believe that a more happy company could have been found in America, large as it is. 'Tis true that an agreeable companion to join me in a glass of punch, to drink a health to absent friends, to speak of by-gone times and speculate on the future, might have made the evening pass more pleasantly, yet I was far from unhappy. To hear the merry joke, the hearty laugh, and lively song among my men, was of itself a source of much pleasure.'[2]

Meanwhile, somewhere off the coast of King William Island, some 500 miles to the north-west, Sir John Franklin's ships were trapped in unmoving pack ice. The bodies of three young seamen already lay in graves on Beechey Island, where the expedition had spent the previous winter.

In April 1847 Rae resumed his explorations, but this time he opted to travel overland. He had bought some sledge dogs from the local Inuit people, and his carpenter, Richard Turner, fashioned two sledges out of battens from the boats. Rae learned how the Inuit repaired their sledges, by soaking snow and chopped moss in water and pressing it onto the runners.

Wearing snowshoes, he took a small party of men and travelled up the west coast of the Gulf of Boothia. He didn't reach the mouth of the Castor and Pollux River, but he went far enough to answer one important question: the land of Boothia itself, once thought to be an island, was actually a peninsula. If there was a missing link in the Northwest Passage, it was not here. Then he turned around, retraced his steps to Repulse Bay, and set out again, this time exploring the Gulf of Boothia's east coast.

This time Rae's progress was excruciatingly slow. Either he and his men were sinking waist-deep in snow, or wading through water up to their knees, or floundering around on ice so glassy that their frozen shoes could find no grip. They pressed on, reaching a point which Rae considered to be within 25 miles of the Fury and Hecla Strait, a sea passage named after the ships of Sir Edward Parry in 1822. Rae had achieved his goal: his surveys had linked the findings of Parry with those of Sir John Ross in 1830. Having mapped over 650 miles of coastline, he turned to make the long haul back to Repulse Bay.

Later, in his account of the expedition, Rae wrote: 'I have… often accomplished between forty and fifty, and, on one occasion, sixty-five miles in a day on snow shoes, with a day's provisions, blanket, axe, etc. on my back; but our journey hitherto had been the most fatiguing I had ever experienced. The severe exercise, with a limited allowance of food, had much reduced the whole party, yet we were all in excellent health; and although we lost flesh, we kept up our spirits, and marched merrily on…'[2]

Had Rae continued his exploration up the west side of the Gulf of Boothia, and trekked across a narrow neck of land that made Boothia a peninsula, he might have gazed across at King William Island where, on the north-western shore, a handwritten note was being inserted into a metal tube and pushed into a cairn of stones. But Rae had no reason to go there, and every reason to turn back. In early September, after an absence of 15 months, Rae and his men returned to York Factory with their mission accomplished.

Sir George Simpson was delighted with Rae's success and offered him a promotion. With spare time on his hands, Rae boarded an HBC ship bound for London, intending to write his report and visit his family in Orkney.

On 1st November 1847, Rae's account of his expedition was published in *The Times*. It was read by Sir John Richardson, a naval surgeon and naturalist. The British public, and in particular Sir John Franklin's wife, Lady Franklin, were becoming anxious for news of the naval expedition which had still not emerged from the Arctic. The ships' whereabouts could only be guessed, but Richardson had been given the task of organising a search and rescue mission. He had been looking for a suitably experienced candidate to join him. When he read Rae's report, he knew he had found his man.

Rae's Arctic journey with Richardson, completed between 1848 and 1849, failed to turn up any evidence of Franklin's fate, and another expedition conducted by Rae between 1850 and 1851 was similarly unsuccessful. Ironically, on a geographical survey mission in 1853, when he believed he had no hope of finding traces of Franklin because it was focused on the wrong area, Rae met Inuit people who offered him news and personal artefacts that had undoubtedly belonged to Franklin and his men.

When they were published in Britain, Rae's findings about Franklin's expedition dealt a devastating blow to his reputation. Ironically, he had also sighted a piece of coastline that represented the sought-after link in the Northwest Passage. This, however, paled into insignificance beside his revelation, unintentionally reproduced in full in *The Times*, that Franklin and his men were dead, but had resorted to cannibalism in a bid to survive. Among the public, a shock wave of horror turned to disbelief and outrage. Lady Franklin, having once pleaded with Rae for help in finding her husband, became his most vengeful critic. Overnight, Rae was transformed from esteemed explorer to social pariah.

It was several years before Rae's account was corroborated by other searchers who went looking for Franklin. His extraordinary accomplishments in surveying large parts of Canada's coastline never received the acclaim they deserved, but they were still revered in the minds of Arctic explorers like Vilhjalmur Stefansson, who, writing in 1913, described Rae as 'a man exact and truthful and in his methods of travel a generation ahead of his time.'[3]

*Today, the settlement at Repulse Bay is known as Naujaat, and it lies exactly on the Arctic Circle, with the Melville Peninsula stretching away to the north.

1 R M Ballantyne, *Hudson's Bay, or Everyday Life in the Wilds of North America* (1848)

2 John Rae, *Narrative of an Expedition to the Shores of the Arctic Sea in 1846 and 1847* (1850)

3 Vilhjalmur Stefansson, *My Life with the Eskimo* (1913)

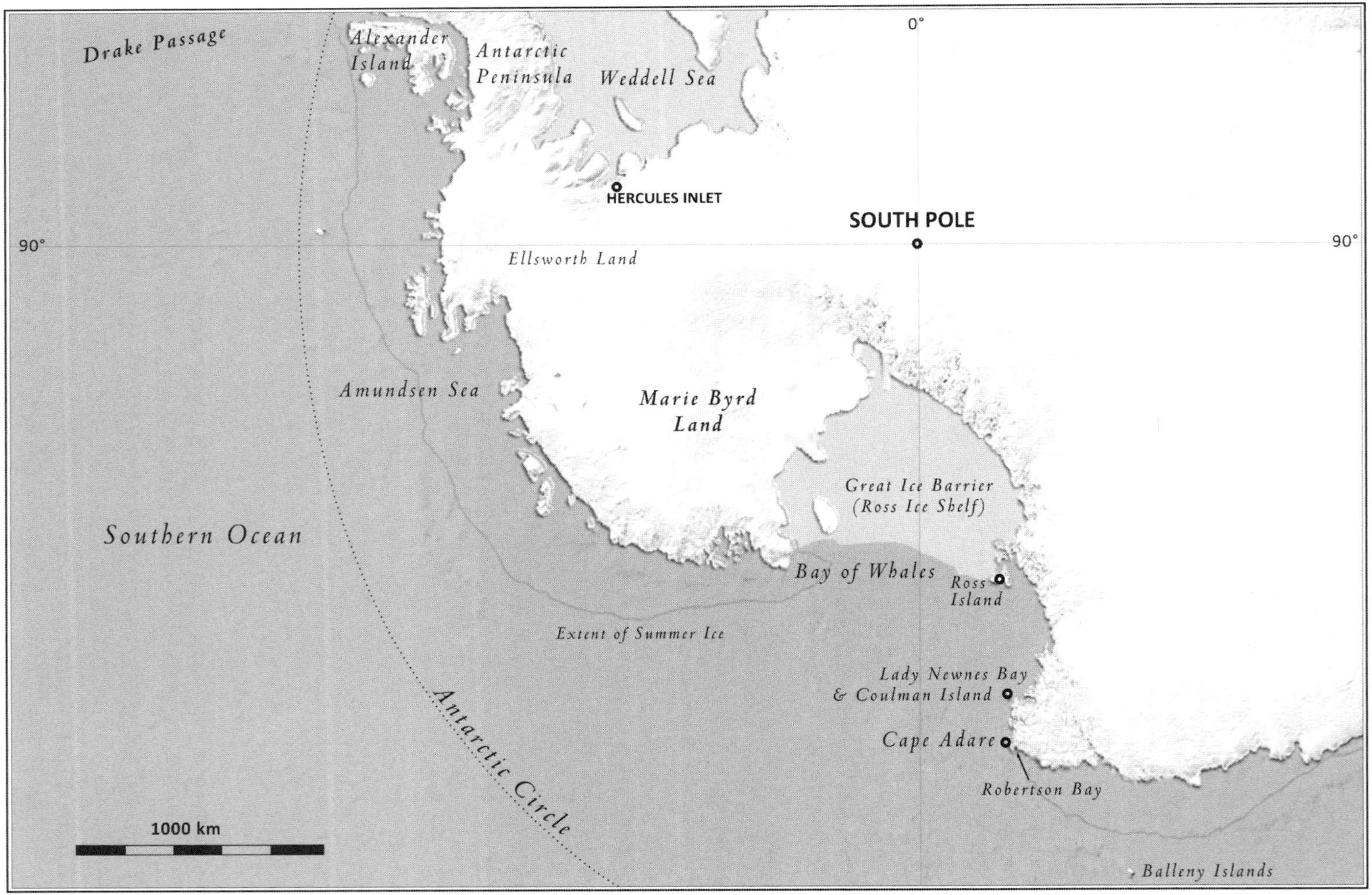
Drake Passage
Alexander Island
Antarctic Peninsula
Weddell Sea
0°
HERCULES INLET
SOUTH POLE
90°
90°
Ellsworth Land
Amundsen Sea
Marie Byrd Land
Great Ice Barrier (Ross Ice Shelf)
Southern Ocean
Bay of Whales
Ross Island
Extent of Summer Ice
Lady Newnes Bay & Coulman Island
Cape Adare
Antarctic Circle
Robertson Bay
1000 km
Balleny Islands

Carsten Borchgrevink

o

Carsten Borchgrevink had always maintained that he was the first person ever to set foot on the Antarctic continent. On 10th February 1900, he might all too easily have become the first man ever to get washed off it.

It was the height of the Antarctic summer, and with much of the Ross Sea free of ice, a couple of Norwegian explorers were having a fine time island-hopping. Having scrambled ashore from a small boat, Borchgrevink and his ship's captain, Bernhard Jensen, were busy collecting rock and plant specimens from a gravelly beach on Ross Island. Somewhere above their heads the dormant volcano of Mount Terror rose to its ice-bound summit. Out in the bay, the *Southern Cross* rode at anchor.

Both of them heard the noise at the same time, a deep roar that seemed to come from somewhere overhead. Borchgrevink glanced up, thinking that a rock face was avalanching. But then, as the noise intensified, he and Jensen turned and looked along the coast to the west, where a glacier flowed out to sea. In fascinated horror, they watched a colossal slab of ice shear away and topple majestically into the sea, churning the water and sending clouds of spray hundreds of feet into the air.

Borchgrevink knew immediately what would happen next. He and Jensen looked desperately around at the narrow beach. At best, it was about four feet above sea level, backed by a near-perpendicular cliff that neither of them could hope to climb. Across the bay, a giant wave had already risen and was coming towards them like a wall. Leaning against the cliff, they grasped handholds of rock and braced themselves for the impact.

Like limpets, Borchgrevink and Jensen clung on through several violent inundations of ice-splintered water. Still in one piece, although 'a good deal knocked about', [1] they were collected up by their comrades on the *Southern Cross*, which had ridden out the iceberg-tsunami without mishap, and the expedition resumed its meandering homeward course.

Born in Christiania (now Oslo) in 1864, Carsten Egeberg Borchgrevink studied at the Royal Saxon Academy of Forestry in Tharandt, Germany, and emigrated to Australia at the age of 24. Initially, he worked as a surveyor in Queensland and New South Wales and then became a teacher of languages and natural sciences at the Cooerwull Academy. But he also had a long-standing interest in the polar regions, and this was reinforced when a fellow Norwegian, Henryk Johan Bull, organised a whaling expedition to the Antarctic. When Bull's ship, the *Antarctic*, called in to Melbourne on the outward journey, Borchgrevink signed on as a deck-hand.

During Bull's two-year expedition, the *Antarctic* visited Tristan da Cunha and several other remote archipelagos before arriving in the Ross Sea. On 24th January 1895, a party of six or seven men, Borchgrevink among them, landed in a small boat at Cape Adare, thereby making the first officially recorded footfall on the Antarctic continent. Borchgrevink subsequently made it clear that he was the first of the party to go ashore.

No one was ever able to confirm Borchgrevink's claim to the 'giant leap' of polar exploration; some of his companions certainly disputed it.* When he returned home, he was more interested in how he could use this status to put together a new expedition, one that he himself would lead. This time, he wouldn't just land on Antarctica: he would spend an entire winter there.

Unable to raise the necessary funds for his expedition in Australia, Borchgrevink travelled to Britain where he found a sponsor in the shape of Sir George Newnes, a wealthy publisher. This allowed him to purchase a former whaling vessel, the *Pollux*, which was both steam-powered and barque-rigged. He replaced her engines and re-named her the *Southern Cross*.

During a lecture to RSGS members in Aberdeen in March 1898, Borchgrevink assured his audience that the *Southern Cross* was 'the strongest vessel afloat except the *Fram*.'[2] This was an impressive claim: *Fram* belonged to the celebrated explorer Fridtjof Nansen, and had been specially designed by Scotsman Colin Archer with a rounded hull that would rise above the pack ice instead of being crushed by it. Archer had also designed the *Southern Cross*. At a similar lecture in Dundee, Borchgrevink revealed that his ship's company would be partly Norwegian and partly British; he added that he would be willing to take sailors from Dundee, and that anyone desiring to sail with him should present themselves to him in person.

> *'The* Southern Cross *was the strongest vessel afloat except the* Fram.'

Perhaps because of his British sponsorship, and the fact that his mother came from an English family, Borchgrevink decided that his expedition would sail under the British flag. Yorkshireman William

Colbeck and Hugh Blackwall Evans from Gloucestershire joined his team of mainly Norwegian scientists and crew. One exception was the physicist Louis Bernacchi, born of a Belgian family who had settled in Tasmania; Bernacchi would later return to the Antarctic with Captain Robert Falcon Scott. Borchgrevink also hired Persen Savio and Ole Must from Finnish Lapland to handle his team of about 70 sledge dogs, which he hoped would propel him to the Magnetic or Geographical South Pole.

The *Southern Cross* sailed out of St Katharine's Dock in London on 22nd August 1898, cheered on her way by a large and enthusiastic crowd of onlookers. Travelling via Hobart in Tasmania, she encountered the first pack ice on 30th December. 'Great was the enthusiasm amongst those on board,' wrote Borchgrevink, 'who saw Antarctic pack for the first time.' [1]

The ship's bows were made of solid oak, 11 feet thick, and their strength was about to be thoroughly tested. From the crow's nest, steering directions were relayed to the engine room far below. After a cry of 'Full speed ahead!', the *Southern Cross* rammed the ice head-on. Borchgrevink described the aftermath: 'Great green blocks of ice rear on end, overturn and plunge against her fortified sides...' [1] Shuddering with every impact, the ship barged her way southwards.

On 14th January 1899, Borchgrevink went up to the crow's nest and sighted land which he identified as 'Balleny Island.'† Thereafter, thickening ice hampered the ship's progress until she was trapped and then lifted up by converging ice sheets. Her timbers groaned under the pressure, and Borchgrevink readied his men in case they needed to abandon ship. But after several ice-bound weeks, the pack loosened and the *Southern Cross* escaped into clear water.

Within days, Borchgrevink was delighted to see the dark rocks of Cape Adare, the place where his feet had first touched dry land four years earlier. Here, he intended to set up his winter camp. He wrote: 'At 11 pm on the 17th February, 1899, for the first time in the world's history an anchor fell at the last *terra incognita* on the globe.' [1]

At first sight, the desolate beach seemed so inhospitable that some of the crew members who intended to overwinter there were filled with foreboding. But they still set to work, landing men, dogs, equipment, coal and provisions. Work commenced on building two timber huts, one for habitation and another for storage; in addition to living quarters, they contained an instrument room, a photographic darkroom and a taxidermy studio. To secure them against the gales, cables were passed over the roofs and attached to ships' anchors embedded in the ground.

On the evening of 2nd March the *Southern Cross* sailed away. She would spend the southern winter in New Zealand, and return early the following year. Behind her she left 10 men whose average age was 26. Borchgrevink wrote: 'We were cut off from all the world, 2,500 miles south of Australia, and all ten of us realised our isolation as the vessel steamed away...' [1]

Luckily there was physical work to do, and Borchgrevink wanted to accomplish as much as possible before the winter brought perpetual darkness. Surveying parties were sent out to climb the surrounding peaks and ridges, and natural history specimens were shot and preserved. In their spare time, ski-jumping contests were organised on a nearby hill, each competitor landing in a soft pile of snow. Meals were based around supplies of tinned meat, fruit and soup, dried vegetables, porridge, bread, jam, cheese, marmalade, tea and coffee; these were supplemented with freshly caught fish, roasted seal and penguin.

On 15th May the sun disappeared for the winter, leaving behind a dwindling red glow on the north-western horizon. Shortly afterwards, a ferocious storm struck the camp, hurling showers of stones at the huts and sending the sledges flying. The huts were soon half-buried in snowdrifts, and an access tunnel had to be dug in the snow. Making daily weather observations in such conditions was a considerable challenge: to avoid being blown away, the meteorologists tied ropes around their waists and crawled on all fours when they went out to their instruments. At night, as the men listened to the tethering cables twanging and whistling ominously in the gale, they half-expected to see their camp lifted into the air like a balloon.

By now, most of the wildlife had headed north, and the men felt increasingly as if they were the only life forms left on the continent. As the days grew inexorably shorter and darker, they resorted to indoor entertainments such as chess, cards and draughts, while listening to a music box whose tunes grew annoying with repetition. Consecutive days of confinement turned into near-claustrophobia, and already they were wearying of each other's company. 'It was a depressing feeling,' recalled Borchgrevink, 'like looking at oneself getting old.' [1]

Louis Bernacchi added with feeling: 'No one who is not young and strong, and devoid of acute sensibilities, should ever spend a winter in the polar regions...' [3]

Although Borchgrevink had great aspirations, he seems to have been uncompromising in his attitude towards others. Antagonism between himself and his men provoked him to draft out a memorandum, warning them that if they ridiculed or criticised him, or opposed his ideas, they would be charged with mutiny. Diplomacy was obviously not one of Borchgrevink's strong points, and meanwhile the tension was exacerbated by alarms. One night, a neglected candle set fire to a bunk and the entire camp nearly went up in flames. Later on, a near-fatal case of carbon monoxide poisoning was caused by draughty holes in the walls being plugged up too efficiently. After these close calls, Borchgrevink collected an emergency supply of tents and provisions in case they needed to make a hasty exit.

The sledge dogs caused their own problems, squabbling among themselves to the extent that several were killed, but then the birth of 16 puppies augmented their numbers. They were, of course, desperate for action, and in late July, as the sun was beginning to return, Borchgrevink harnessed 30 of them to two sledges and took off to explore the coast around Cape Adare with Persen Savio, Ole Must, and scientific assistant Anton Fougner. The dogs proved almost impossible to control, and it was largely thanks to the hunting skills of Savio and Must, who captured and killed a seal, that the party survived.

Undeterred, Borchgrevink set out on a longer, more ambitious trek with Must and Savio, but almost immediately they were caught in a severe storm. As the temperature sank to -40°C (-40°F), their flimsy silk tent was lined on the inside with ice crystals, and their reindeer-hide sleeping bags froze solid. Borchgrevink wrote: 'For three nights and for three days we had to take turn in standing on all fours to prevent being smothered by the pressure of the snow.'[1] Again, when the storm eased Savio helped save the situation by constructing a new tent out of sealskins and bags stretched over the sledges. In Robertson Bay, the little party explored the shore of an island which Borchgrevink named Duke of York Island, after the second son of King Edward VII who would eventually become King George V. Then they turned back towards their camp at Cape Adare.

Ever since the expedition arrived in the Antarctic, it had been increasingly obvious that Nikolai Hanson, the zoologist and taxidermist, was not in the best of health. On 14th October he died of an unidentified illness, possibly an intestinal disorder, and he became the first person to be buried on the Antarctic continent. Dynamite had to be used in order to excavate a grave on the frozen summit of Cape Adare. Just before he died, Hanson was delighted to be told that the penguins were beginning to arrive for the summer, and one of the newcomers was brought to his bedside for him to see.

With the return of the penguins, a supply of fresh eggs and penguin meat was assured. The first eggs were gathered in early November, and soon the men's thoughts turned towards Christmas. Conviviality might have been lacking, but Borchgrevink painted a cosy picture: 'Christmas Eve was celebrated by speeches, toddy, extra rations of biscuits, and a longing for home. On Christmas Day we had tinned plum-pudding, and Mr Evans, in honour of the occasion, baked cakes; and Mr Ellefsen [the cook] even surpassed himself in the excellence of his cooking.' [1] According to Bernacchi, however, 'Christmas Day was hopelessly dull, for an appalling storm blew all day... mental oppressiveness and silence reigned, so we were scarcely delirious with joy...' [3]

On the morning of New Year's Day, 1900, the men turned their shirts inside out for good luck and began to scan the horizon anxiously for their returning ship. There was open water as far as the eye could see, but they had no way of knowing the fate or the whereabouts of the *Southern Cross*. After days of waiting, prospects began to look grim. To Bernacchi, 'it seemed as if some accident had befallen the ship and we were doomed to remain upon that land of desolation for another year, perhaps two, and that was an incubus too dreadful to dwell upon.' [3]

Towards the end of January, Ellefsen noticed a figure approaching across the ice. At first he thought it was a particularly large penguin, but when it drew closer he recognised Captain Jensen of the *Southern Cross*, carrying a sack of mail on his back.

All the other men were asleep, but they were rudely and happily awakened by Jensen throwing his bag down in the hut with a hearty cry of 'Post!' 'With what hysterical joy,' wrote Bernacchi, 'we tumbled out of our bunks to welcome him.' [3] Letters from loved ones were eagerly opened, along with newspapers telling of developments in the outside world. 'We heard for the first time,' wrote Borchgrevink, 'about the war in the Transvaal; about wonderful discoveries in telegraphy, and found how many alterations of conditions one year might cause.' [1]

With the *Southern Cross* anchored in the bay, the shore party prepared to vacate the small huts that they had called home for just under a year. There was no wistfulness at departure; Louis Bernacchi, for one, expressed a heartfelt wish that he might never pass another 12 months in similar conditions. The men paid their last respects at Hanson's grave and boarded the ship, leaving behind some stores and a quantity of coal that might benefit some future expedition.

The *Southern Cross* proceeded along the coast, with Borchgrevink naming fjords, promontories and bays as they passed by. (Some, such as Newnes Glacier and Lady Newnes Bay, are still in use.) Landings were made on Possession Island, Coulman Island and Franklin Island to allow the scientists to collect specimens and make observations. A few days after Borchgrevink and Colbeck survived the iceberg-tsunami on

Ross Island, an inlet was found in the Great Ice Barrier (now the Ross Ice Shelf) which was tentatively explored. Here, Borchgrevink, Colbeck and Savio landed on the Barrier and travelled south for about 10 miles, reaching a latitude that Borchgrevink calculated to be 78°50'S: certainly the farthest south ever reached by man. This inlet proved to be the opening that led future explorers ever closer to the Pole. Sir Ernest Shackleton named it the Bay of Whales.

Overnight on Sunday, 1st April, the *Southern Cross* put into Half Moon Bay, New Zealand, where there was a telegraph office. Borchgrevink composed a message to Sir George Newnes, and waited for the operator to transmit it. 'Suddenly,' he wrote, 'the clerk turned to inform me that some very distant official desired to congratulate me on my safe return, and wished very much to know whether I had found any people "down there".' [1]

Borchgrevink sailed home to Britain, confident that he would be greeted as a returning hero. Unfortunately, that was not the case. A new, government-funded expedition, largely made up of Royal Navy personnel, was about to set out for the Antarctic under the command of Robert Falcon Scott. Despite the fact that he had sailed under the British flag, Borchgrevink's privately funded expedition received a lukewarm homecoming. Public interest, too, was diverted by the ongoing war in South Africa. Borchgrevink was, however, made a Fellow of the Royal Geographical Society and in Norway he became a Knight of the Royal Norwegian Order of St Olav.

When Borchgrevink returned to Scotland to give lectures in November 1900, RSGS awarded him the Silver Medal and Honorary Fellowship of the Society. Before his lecture in Dundee, he was introduced as an 'old friend' who was sure of a cordial welcome, and he responded with the touching if slightly exaggerated claim that 'he felt the success of the last expedition under his control was largely due to the enthusiastic support which he received in Scotland, so near by heart, habits and feeling to the Norwegian nation.' [4]

During his talk in Aberdeen, which was also attended by Captain Jensen of the *Southern Cross*, Borchgrevink expressed the hope that the forthcoming Scottish National Antarctic Expedition, currently being planned by William Speirs Bruce, 'might benefit by their experience, and that they might find the difficulties in some way minimised.' [5]

Borchgrevink seems to have possessed an excellent opinion of himself, a quality that did not endear him to everyone, in particular his own men. He attracted serious criticisms, including poor leadership, ignoring the work of his scientists, and the journalistic style of his subsequent book, *First on the Antarctic Continent*. But he lost only one man, apparently to natural causes.

After Scott's second expedition (1910-1913) and the gruelling experiences of his Northern Party at Cape Adare, it was clear that the *Southern Cross* expedition had been undervalued. In 1911, Roald Amundsen anchored his ship in the Bay of Whales before embarking on his successful trek to the South Pole. 'We must acknowledge,' he wrote, 'that by ascending the Barrier, Borchgrevink opened a way to the south, and threw aside the greatest obstacle to the expeditions that followed.'[6]

* A sealer, John Davis, challenged all these claims, saying that he had stepped onto the Antarctic Peninsula in 1821.

† The Balleny Islands are in fact an archipelago.

1 C E Borchgrevink, *First on the Antarctic Continent: Being an Account of the British Antarctic Expedition, 1898-1900* (1901)

2 *Aberdeen Free Press*, 24th March 1898

3 Louis Bernacchi, *To the South Polar Regions: Expedition of 1898-1900* (1901)

4 *The Dundee Courier*, 17th November 1900

5 *The Scotsman*, 16th November 1900

6 Roald Amundsen, *The South Pole: An account of the Norwegian Antarctic expedition in the Fram, 1910-1912* (1912)

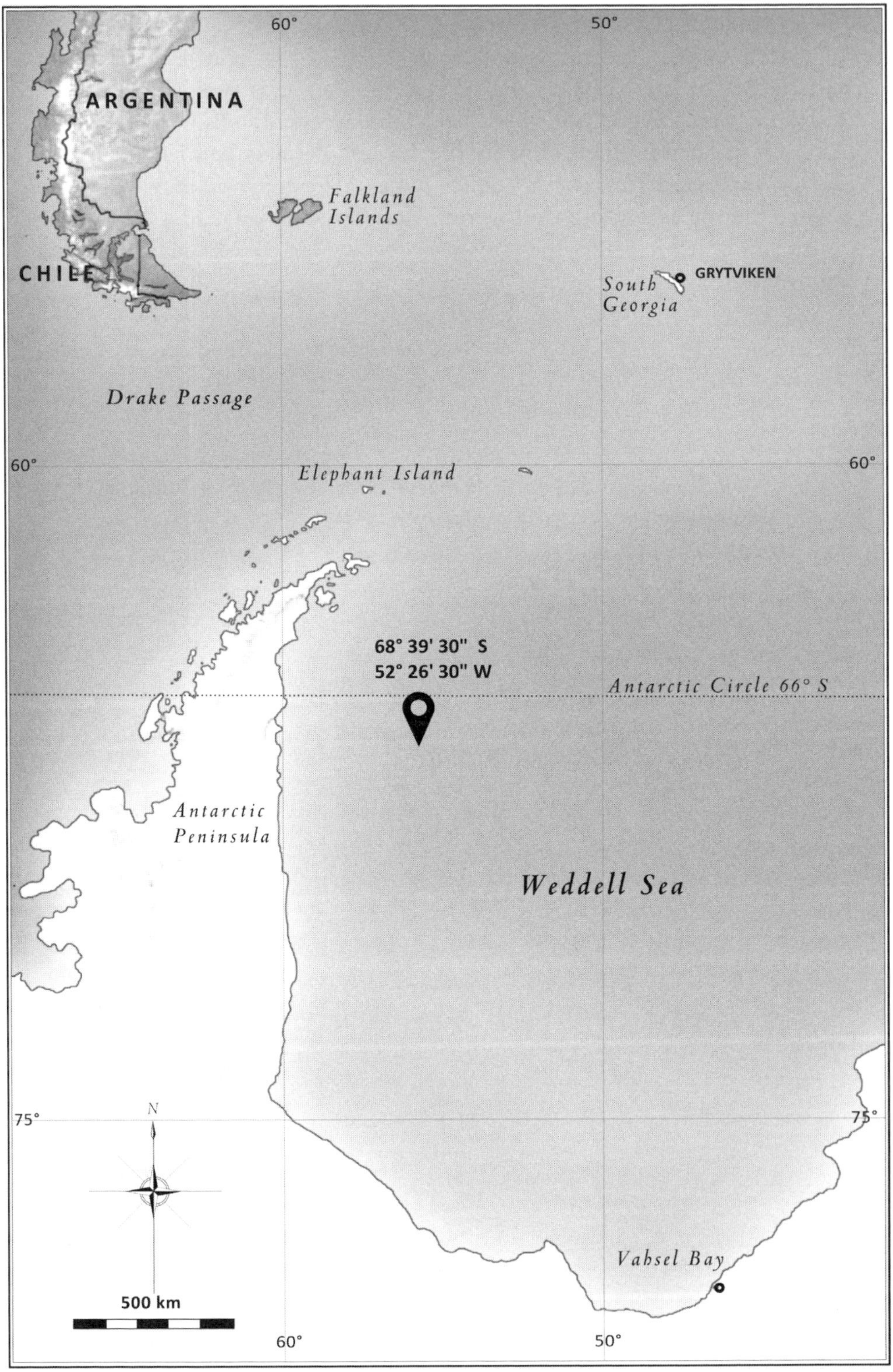
60°
50°
ARGENTINA
Falkland Islands
CHILE
GRYTVIKEN
South Georgia
Drake Passage
60°
Elephant Island
60°
68° 39' 30" S
52° 26' 30" W
Antarctic Circle 66° S
Antarctic Peninsula
Weddell Sea
N
75°
75°
Vahsel Bay
500 km
60°
50°

Sir Ernest Shackleton and the Endurance22 Expedition

o

'I'll be blowed if I want to see any more ice as long as I live!' wrote Reginald James in his diary. 'I expect most of us feel the same, but we manage to conceal it fairly well nevertheless.'[1]

It was late October, 1915. *Endurance*, the ship of Sir Ernest Shackleton's Imperial Trans-Antarctic Expedition, had been beset in pack ice for nine months. Pressure ridges thrown up by jostling ice floes repeatedly shook, squeezed and twisted the stricken ship, retreating briefly only to resume the attack with renewed vigour. On one occasion she had been lifted and then thrown abruptly onto her side, flinging dogs, kennels, sledges and stores across the deck, and settling at a 30-degree angle which forced the men to brace their legs against battens as they sat to eat their dinner in the wardroom.

When the *Endurance* righted herself, the men, ever hopeful, started to scan the pack for signs of a lead. Ice debris was cleared from the ship's rudder, her boilers were filled, and the fires were banked up in readiness for departure under steam. But all was in vain. The onslaught began again, and a particularly dramatic crash signalled that the sternpost, which supported the rudder, had almost been ripped out. With the ship leaking dangerously, Shackleton gave the order to raise steam and start the pumps, but the water continued to rise. In pitch darkness, and fearing that creaking timbers would collapse on them at any moment, Hubert Hudson, Lionel Greenstreet and Frank Worsley, the ship's captain, dug down through the coal bunkers to reach the bottom of the bilge pump, which was blocked with ice, and cleared it with a blowtorch.

Hour after hour, more pumping was done manually in shifts, while the carpenter, Harry 'Chippy' McNeish, worked waist-deep in water to build a cofferdam across the stern. It was desperate, exhausting work. As if to underline the futility of it all, a group of Emperor penguins arrived and regarded the scene intently before throwing back their heads and uttering a disturbingly eerie cry. Worsley wrote: 'I have never, either before or since, heard them make any sound similar to the sinister wailings that they moaned that day.'[2] The pumping continued, but he knew in his heart that his ship was doomed.

Published early in 1914, the official prospectus for the Imperial Trans-Antarctic Expedition, drawn up at the Expedition's offices in New Burlington Street, London, describes Shackleton's daring plan to cross the Antarctic continent via the South Pole but makes no direct mention of the risks involved. This is hardly surprising: Shackleton was looking to attract the interest of investors, after all. Acknowledging the achievements of earlier pioneering expeditions in Antarctica, including that which claimed the life of Captain Robert Falcon Scott and four of his men, Shackleton presented his new adventure as a natural sequel of great historic and scientific value.

If all went to plan, Shackleton and five companions, travelling with 100 sledge dogs and two motor sledges, would strike inland from Vahsel Bay on the Weddell Sea and reach Ross Island in McMurdo Sound five months later. During the crossing of the continent, a journey estimated at 1,800 miles (2,897 km), important observations would be made in the fields of topography, geology and glaciology; continuous magnetic recordings would be taken, and meteorological conditions noted. For the first half of the journey in particular, the terrain was still largely uncharted: 'Every step,' explained Shackleton, 'will be an advance in geographical science.'[3] He added, with unwitting prescience, that his expedition aimed to accomplish 'the greatest Polar journey ever attempted.'[3]

For this ambitious undertaking, Shackleton had acquired two ships. One, the *Aurora*, would be stationed in the Ross Sea to await the arrival of the crossing party, and her crew would lay food depots inland to assist their progress. The other ship, based in the Weddell Sea, would deposit 14 men on shore and then depart to make surveys up and down the coast. Shackleton wrote: 'She is barquentine rigged, and has triple-expansion engines giving her a speed under steam of 9 to 10 knots... She is of about 350 tons, and built of selected pine, oak and greenheart.'[3]

Her original name was *Polaris*, and beneath the letters on her stern she carried a five-pointed star. She had been specially built in Sandefjord in Norway for shipowner Lars Christensen, who intended to run 'polar safaris' to Spitsbergen and Greenland in partnership with the Belgian polar explorer Adrien de Gerlache. But their scheme collapsed before any of these tours could set sail, leaving a custom-designed, ice-strengthened ship in need of a new adventure. In March 1914, Shackleton purchased her for £13,000. He changed her name to *Endurance*, honouring his family motto, *Fortitudine Vincimus*: 'By Endurance We Conquer.'

Shackleton had been on two previous Antarctic expeditions. On the first occasion, he'd served as Third Lieutenant on Scott's ship RRS *Discovery*, and together with Scott and Edward Wilson he had made a gruelling three-month sledge journey, getting within about 460 miles of the South Pole. On his second visit to the Antarctic he'd led his own expedition on board a former sealing vessel, *Nimrod*, and had attained a new 'Furthest South'; his expedition had also made the first ascent of Mount Erebus and located the South Magnetic Pole.

Hailed in Britain as a returning hero and knighted by King Edward VII, Shackleton still cherished an unfulfilled dream. He wasn't done with the Antarctic, or perhaps the Antarctic wasn't done with him. In his account of the *Nimrod* expedition, he wrote: 'Men go out into the void spaces of the world for various reasons. Some are actuated simply by a love of adventure, some have a keen thirst for scientific knowledge, and others again are drawn away from the trodden paths by the "lure of little voices," the mysterious fascination of the unknown.' [4] While the South Pole had been attained by the Norwegian explorer Roald Amundsen in 1911, the crossing of Antarctica from sea to sea would 'secure for the British flag the honour of being the first carried across the South Polar Continent.' [3]

On 8th August 1914, the *Endurance* sailed out of Plymouth and away from British shores. With the outbreak of the First World War just days earlier, Shackleton had delayed briefly to offer ship and crew to the service of the Admiralty, but the offer was declined with a one-word telegram: 'Proceed.' Shackleton wrote: 'We are leaving now to carry on our white warfare, and our farewell message to our country is that we will do our best to make good.' [5]

But at the whaling station at Grytviken, South Georgia, the last port of call before the Antarctic continent, the whaling captains counselled caution. Ice conditions that winter had been severe, and the pack ice had extended further north than any of them could recall. Heeding their advice, Shackleton lingered until 5th December before setting sail. Just three days later, the *Endurance* encountered the outskirts of pack ice.

The Weddell Sea covers an area of around 1.7 million square miles (4.4 million square km). In September, the depth of the austral winter, its entire surface is covered by sea ice to an average depth of five feet, and air temperatures can plummet below -34°C (-29°F). In summer, when the ice retreats and temperatures can rise to a comparatively balmy 4.4°C (40°F), a current known as the Weddell Sea Gyre circulates loose floes, pack ice and huge icebergs around a natural basin that is enclosed on three sides by ice shelves.

It was into this grim carousel of monsters that the *Endurance* nosed her way, weaving a cautious route south. If an obstructive floe was no more than about three feet thick, it was rammed with the ship's bows in an attempt to split it. Frank Hurley, the photographer, explained the procedure: 'Worsley selects a point of attack that promises a line of weakness—"Full speed ahead" is rung down to the engine room. The engines throb and the ship quickly gathers speed, hastening forward to the charge... There is a mighty collision as if we have run on a reef. The ship is brought up all standing. Her massive cutaway bows rise up on the ice nearly clear of the water. Then as if dulled by the violence of the concussion, she slides slowly back, reeling from side to side.' [6]

But the further south they progressed, the harder the task became. On 18th January 1915, with Vahsel Bay still 80 miles (128 km) distant, ice closed around the *Endurance* and defied the men's best efforts to break her free. Hurley wrote: 'As far as the eye could reach in every direction no water was visible. The ship was a helpless atom, locked in, and drifting helplessly with the pack.' [6] As winter tightened its grip, the *Endurance* was carried slowly northwards with the Weddell Sea gyre.

'The ship was a helpless atom, locked in, and drifting helplessly with the pack.'

In the dwindling daylight, the men kept their spirits up by playing football and organising dog races on the ice; during the darkest months, spare time below decks was occupied with cheerful singsongs around Leonard Hussey's banjo, and games of cards, chess and draughts.

A programme of light-hearted entertainment was put on to mark Midwinter's Day; among the performers was physicist Reginald James, who was hailed by Worsley as 'the gem of the evening' with his reading of a spoof academic paper that was comic for being utterly unintelligible. Although storms howled in the rigging, laughter and hope were kept alive: next season, they might still attempt the crossing.

The Antarctic spring, however, delivered savagery instead of salvation. Loosening and colliding floes set up a bombardment that shook the ship from mast to keel, and squeezed her hull until it groaned as if in pain. After the sternpost was smashed, nothing could be done to save her. Pumping and dam-building would merely stave off the end.

On 27th October the men ate their dinner in silence, listening to the clock ticking on the wardroom wall. At 5 pm Shackleton gave the order to abandon ship. As the last man to leave, he wrote: 'It was a sickening sensation to feel the decks breaking up under one's feet, the great beams bending and then snapping with a noise like heavy gun-fire.' [7] He hoisted the blue ensign so that she would go down with her colours flying. Then he climbed overboard to join his men on the floe.

Dogs, sledges, lifeboats and a quantity of provisions had already been landed on the ice. There was no chance of rescue. To the men gathered around him, Shackleton quietly explained how they were going to get home: they would march over the pack ice to Paulet Island, 346 miles (557 km) to the north, where he knew there was a hut containing stores left behind by the Swedish explorer Otto Nordenskjöld in 1903.

But dragging such enormous loads over unstable ice proved impossible. Instead, Shackleton established camp a few miles away from the *Endurance*, with the idea of drifting on a floe and launching the lifeboats when open water appeared. More supplies were salvaged from the abandoned ship, and Hurley dived through icy water to rescue precious photographs and cine film.

On the evening of 21st November, a call rang out from Shackleton, who was keeping watch from a vantage point: 'She's going, boys.' The men clambered up to gaze across at their ship. Her stern rose into the air and then she slid, bows first, under the water. Ice quickly closed up the pool where she had been.

As a navigator, Worsley prided himself on his skill with a sextant and chronometer. Unfortunately, for three days before the *Endurance* sank, and for 19 hours afterwards, thick cloud prevented him from taking a reading from the sun. But under a clear sky at noon the next day, he estimated her last position and jotted down these coordinates: 68°39'30"S, 52°26'30"W.

In February 2018, having obtained special permission to access the archives of the Imperial Trans-Antarctic Expedition held at the Scott Polar Research Institute in Cambridge, polar geographer Dr John Shears found himself staring in silence at the figures pencilled into Worsley's diary.

For over a hundred years, it had been thought that the *Endurance* was lost forever. Although the dream of finding her had long tantalised explorers, scientists and historians, the hostile nature of the Weddell Sea always proved to be too great a challenge. Even supposing that a vessel could reach the precise spot where the *Endurance* sank, the ensuing search would have to be conducted at a depth of over 10,000 feet (3,050 m) across an uncharted sea floor.

But in the 21st century, our knowledge of the Weddell Sea has increased; significant advances have been made in satellite remote sensing, subsea technology, and the capabilities of icebreaking ships. The Weddell Sea Expedition, organised and funded by the Flotilla Foundation, was announced in April 2018; its two main aims were to study the waters around Antarctica's Larsen C Ice Shelf, whose stability was causing scientists concern, and to locate the wreck of the *Endurance*.

As expedition leader, John Shears worked alongside chief scientist, Professor Julian Dowdeswell, and subsea offshore manager, Claire Samuel. Director of exploration was Mensun Bound, a marine archaeologist. The ship chosen to carry the expedition was the SA *Agulhas II*, a South African polar research and logistics vessel. Captain Knowledge Bengu had commanded the ship since 2013, and he was joined by Captain Freddie Ligthelm, an expert on the pack ice of the Weddell Sea. From making regular annual voyages to a South African research base in the Antarctic, most of the ship's 45-strong crew had considerable experience of operating in this notoriously hostile region.

Departing from Penguin Bukta in East Antarctica in January 2019, the SA *Agulhas II* sailed first to the Larsen C Ice Shelf. Here, imagery of the sea floor collected by an AUV (Autonomous Underwater Vehicle) showed a pattern of sedimentary ridges dating back to the end of the last ice age, about 12,000 years ago; scientists interpreted these ridges as the impressions of a floating glacier that was repeatedly sinking and rising with the tides, and deduced that it had been retreating at a rate of roughly six miles per year, 10 times faster than the rate observed by satellites today.

But the search for the *Endurance* was delayed by a technical issue with an ROV (Remotely Operated Vehicle). Fixing the problem cost precious time, and then disaster struck. When the ship reached the wreck site - an extraordinary achievement in itself - an AUV that had been sent down to investigate the sea bed failed to make contact with the ship. After an intense 40-hour search in deteriorating conditions, the team was forced to accept that it was lost. With winter fast approaching, the expedition was called off.

'It was,' wrote John Shears, 'a hard and bruising defeat to take.' [8] He thought it would be years before another attempt was made to find Shackleton's ship.

But in August 2020, John received a call from Donald Lamont, a former Governor of the Falkland Islands and then Chairman of the Falklands Maritime Heritage Trust. John learned that the Trust was planning a major new enterprise in the Weddell Sea. There would be another search for the *Endurance*.

John was appointed leader of the Endurance22 Expedition, while the subsea manager was Nico Vincent, a marine engineer renowned for his skill in solving deepwater challenges. Nico reviewed the operating philosophy of the previous expedition and made some key changes to the subsea equipment. As chief scientist, Dr Lasse Rabenstein headed an international team of sea ice researchers and engineers, and Mensun Bound returned as director of exploration. Joining them was the TV historian Dan Snow, and a team of film-makers led by Natalie Hewit. Once again, Captain Knowledge Bengu and ice-master Captain Freddie Ligthelm would be steering SA *Agulhas II* through the ice-choked waters of the Weddell Sea.

When the SA *Agulhas II* sailed out of Cape Town on 5th February 2022, she carried two new Sabertooth AUVs, manufactured by Saab in Sweden. Crucially, these vehicles could operate in deep water either as autonomous vehicles or by manual control via a fibre-optic tether. Using the tether meant that survey data could be received by the subsea team on the surface as soon as it was collected, and if something went badly wrong, the vehicle could be recovered from the sea bed. John had asked a team of experts to reassess the previous expedition's search box, taking into account a wide range of historical information along with possible chronometer errors and ice movements; the result was a rectangular area measuring 15 by eight nautical miles, offset to the east and south of Worsley's coordinates, and in a slightly different position to the one used in 2019.

On the third day of the voyage a cyclone arrived, sending giant waves crashing over the rear deck and forcing Captain Knowledge Bengu to divert the ship's course away from the centre of the weather system. Ice pilot Freddie Ligthelm observed: 'It makes you think how Sir Ernest Shackleton and his men must have felt in 1915 aboard a wooden, underpowered vessel venturing in this hostile environment... In those days, the ships may have been made out of wood, but the men were made out of steel.' [8]

As the ship approached the search box, John organised a helicopter reconnaissance flight and was deposited on a large, smooth floe with Lasse Rabenstein and four other scientists. 'As I walked across the ice,' he wrote, 'the cold wind blew hard into my face, and I realized in amazement I was perhaps tracing the footsteps of Shackleton and his crew.' [8] With an ice drill they measured the floe's thickness, and installed an ice monitoring buoy to provide data on the speed and direction of drift. It was time to go looking for the *Endurance*.

Just after midnight on 20th February, they thought they had found her. In the subsea operations room, the sonar screen showed an anomaly which caused widespread celebration. But when the corresponding video was inspected, all that could be seen were several long humps in the seabed and a small piece of wood poking out of the mud. It wasn't the *Endurance*, but it could be debris from the wreck.

That was the slim hope that the expedition team clung to, knowing that they were running out of time.

Following discussions with the ship's captain, John and Nico requested, and were granted, a 10-day extension to the expedition. As February turned to March, the days shortened and the wind chill dropped to -25.3°C (-13.5°F); despite being swathed in layers of warm clothing, team members working on deck were forced to take shelter indoors at half-hour intervals. By 5th March, 81% of the search box had been covered. It was the largest subsea survey ever conducted by AUVs under drifting ice, but only three more days remained before the hunt would have to be abandoned. A huge floe had drifted over the search box, making life more difficult for the subsea team, who continued to work around the clock. As Nico Vincent remarked, 'How can you be part of Shackleton's story and give up?'[9]

'How can you be part of Shackleton's story and give up?'

After the Sabertooth AUV set off on her 30th dive, John disembarked from the ship with Mensun Bound and together they walked over to a nearby iceberg. Knowing that failure was likely, they talked about how they might break the news to the rest of the team and the outside world. As they began to walk back to the ship, John turned to Mensun and said with a smile: 'Today will be a good day. I think she's beneath our feet!'[8]

At about the same time, on the sonar screen in the subsea control room, an extraordinary image was beginning to appear. This time, the data clearly showed a wreck.

Trying hard to rein in his excitement, Nico Vincent asked for the AUV's onboard camera to be switched on. Her batteries were dangerously low, but what everyone needed now was visual confirmation. Obediently, the AUV flew low over the ocean floor, transmitting images as she went. The subsea team waited, scarcely breathing.

Out of the darkness, the *Endurance* appeared. She sat upright on the sea bed, still clad in her original black paint. Her masts, stays, ropes, pulleys, rigging and funnel lay haphazardly about the deck. She looked intact and majestic. 'She was,' said Nico, 'beyond our wildest dreams.'[8]

After this moment of pure and utter joy had been shared with everyone on board, the team worked hard to survey the wreck in the two days that were left to them. The Sabertooth's laser scanner produced a 3D digital model of the wreck at 1 mm resolution, and its digital camera took 25,000 high-quality photographs. Video footage showed the ship's wheel still in place on the aft well deck, while on the stern the five-pointed star and nine letters of her iconic name gleamed in the searchlight.

She had been found at a depth of 9,869 feet (3,008 m), just 5.4 nautical miles southeast of the position estimated by Frank Worsley. Because of the deepwater conditions of the Weddell Sea, she is in an astonishing state of preservation. Artefacts can be seen scattered on deck - plates, pots, mugs, rolled-up pieces of lino flooring, even a leather boot. 'It is our hope,' wrote Nico Vincent, 'that the 3D digital model becomes the lasting testimony of our expedition and awakens a wide audience and new generations to Shackleton and the *Endurance* story.' [8]

Incredibly, Shackleton's ship had been found on the exact day, one hundred years ago, that he had been laid to rest in the little graveyard at Grytviken in South Georgia. He had been taken ill and died on board another ship, the *Quest*, at the beginning of his third expedition (and fourth visit) to Antarctica.

At Shackleton's graveside on 11th March 2022, the Endurance22 Expedition team stood with Captain Knowledge Bengu and other members of the ship's crew and paid their respects to 'The Boss'. They read poems by Browning and Kipling, both favourites of Shackleton, and shared hopes for the return of peace to the world; speaking from his heart as a ship's captain, Knowledge told him: 'We have found your baby.'

Throughout Shackleton's expedition - the sinking of the *Endurance* and the desperate aftermath, which involved an 800-mile (1,287-km) open boat journey, a near-superhuman trek over the mountains of South Georgia, and the successful rescue of a 22-strong party marooned on Elephant Island - his abiding concern was for the safety and welfare of his men. After the quiet ceremony, John reflected on the spirit of friendship and cooperation that had sustained him and the rest of the team during their own adventure in the Weddell Sea. The Endurance22 Expedition, he wrote, 'was a glorious symbol of what can be achieved when people trust each other and work together to meet the greatest of challenges.' [8]

[1] R W James, diary (20th October 1915)

[2] Frank Worsley, *Endurance: An Epic of Polar Adventure* (1931)

[3] Prospectus of the Imperial Trans-Antarctic Expedition (1914)
[4] Sir Ernest Shackleton, *The Heart of the Antarctic* (1909)
[5] *The Daily Chronicle*, 29th October 1914
[6] Frank Hurley, *Argonauts of the South* (1925)
[7] Sir Ernest Shackleton, *South* (1919)
[8] John Shears and Nico Vincent, *Endurance: The Discovery of Shackleton's Legendary Ship* (2024)
[9] National Geographic Documentary Films, *Endurance*, directed by Jimmy Chin, Natalie Hewit, Elizabeth Chai Vasarhelyi (2024)

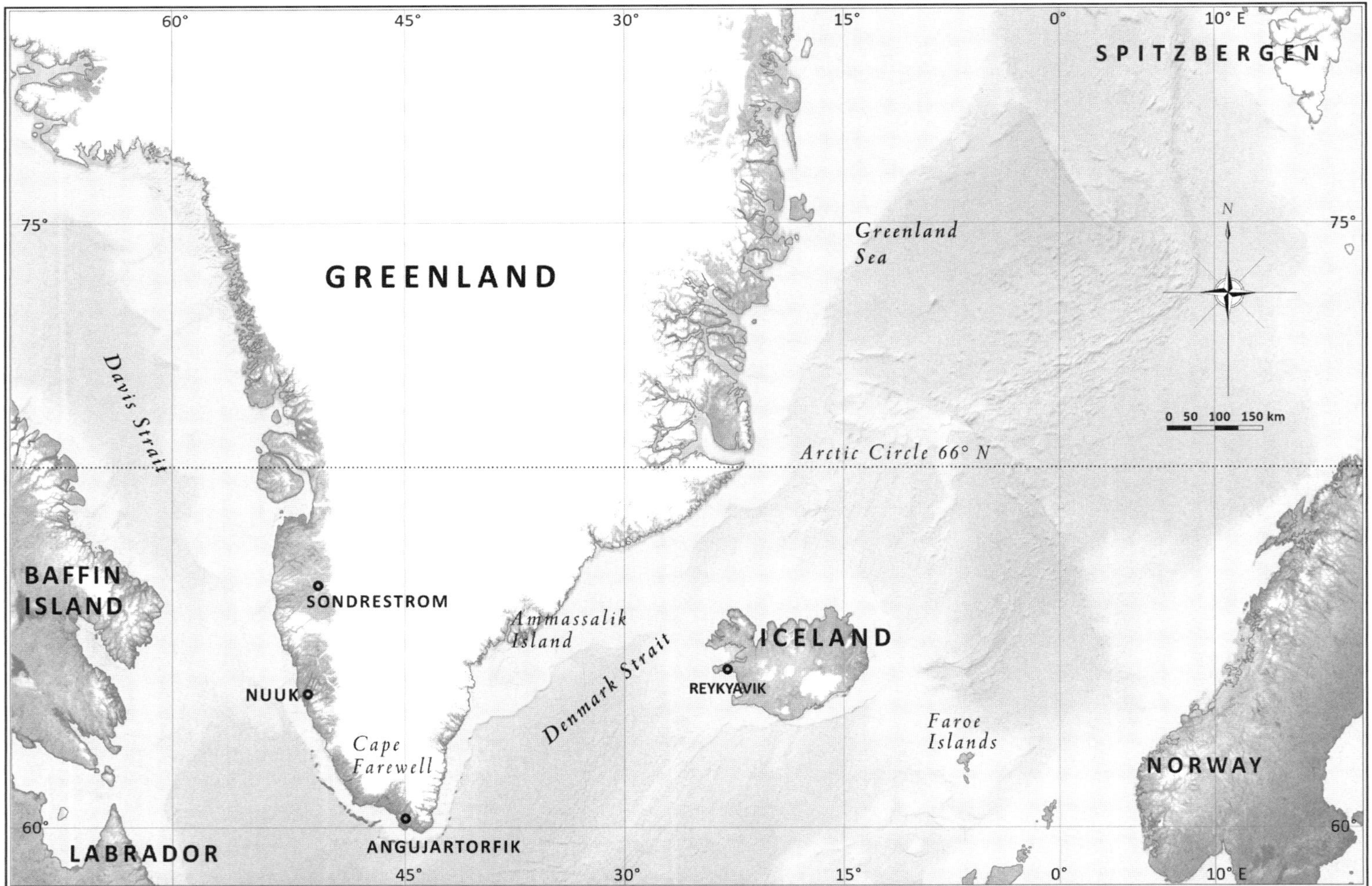

SPITZBERGEN
GREENLAND
Greenland Sea
Davis Strait
Arctic Circle 66° N
BAFFIN ISLAND
SONDRESTROM
Ammassalik Island
ICELAND
Denmark Strait
REYKYAVIK
NUUK
Cape Farewell
Faroe Islands
NORWAY
LABRADOR
ANGUJARTORFIK
N
0 50 100 150 km
60°
45°
30°
15°
0°
10° E
75°

Myrtle Simpson

Myrtle Simpson is remembering the time when she took her baby son to Spitsbergen. Her husband, Hugh, was a pathologist at Glasgow University, and all through 1959 she had been helping him to plan a scientific expedition. Hugh's intention was to study the effect of prolonged daylight on the adrenal gland in people who lived north of the Arctic Circle. Myrtle was tremendously excited, but they were not long into the preparations when she became pregnant. Her baby would be born before they were due to leave.

In 1960, taking a newborn baby to the Arctic for six months was not an idea that many women would have considered. But Myrtle was different. 'I thought, well - stay at home! - I was jolly well not going to do that! So I thought about all the women who do travel all the time, babies or not, and of course that takes you north.'[1]

'I thought, well - stay at home! - I was jolly well not going to do that! So I thought about all the women who do travel all the time, babies or not, and of course that takes you north.'

And she was absolutely right. In Norway, where they stopped on their way to Spitsbergen, a practical solution presented itself. A family of Sámi reindeer herders, camping by a lake, bartered a traditional *komse* or cradle for a bottle of Drambuie. In this cosy nest, the four-month-old Robin was quickly ensconced. Myrtle strapped it onto the frame of her rucksack while she was walking, and she found it also sat well on a sledge or a canoe. 'It's completely windproof,' she explains. 'It's reindeer skin with the fur on the inside, so babies just love it.'[1]

When Robin learned to crawl, he did so over soft tussocks of alpine flowers that carpeted the tundra, and he cut his teeth on strips of rawhide that lashed the sledge. Myrtle and Hugh were experienced travellers; for them, the most important thing was that they were all together. In the years that followed, with a growing family, they made similar expeditions to remote parts of Iceland and the rainforests of Surinam.

Then an even more daring adventure was suggested: an expedition on skis across the Greenland ice-cap. This time, Myrtle and Hugh would be travelling light. There was no room for the children, and even if they could come, the risks were too high. Blizzards, rock walls, crevasses, and hundreds of miles of ice made it an unthinkable proposition. Robin, now aged five, pleaded that he was good at skiing, but it was no use. He and his younger siblings, Bruce and Rona, had to stay behind.

In June 1965, when Myrtle stood outside her tent and gazed out at the vastness of the Inland Ice, the memory of parting from her children was still raw in her mind. But in six weeks' time, a friend would bring them out to the west coast of Greenland. Myrtle's consolation, and the impetus that would drive her forward, was the prospect of seeing them when she got to the other side.

Greenland's ice sheet covers some 660,000 square miles (1.7 million square km), and is the second largest body of ice on the planet (the largest being in Antarctica). It occupies about 80% of the surface of Greenland, and towards the centre it reaches a maximum thickness of 1.9 miles (3 km). But at the edges, it disintegrates into a splintered mass of glacier ice, with crevasses and rock walls to negotiate. Up these considerable obstacles, Myrtle, Hugh and their companions Bill Wallace and Roger Tufft hauled two 400-pound (181-kg) sledges, laden with provisions.

It was back-breaking work, and the sun reflecting off the ice burned their faces. Often they had to lighten a sledge by offloading items and going back for them later. Until they reached the plateau, their skis were useless. Myrtle noticed the paw marks of an Arctic fox crossing her path, and her spirits lifted when a snow bunting fluttered up and burst into song. The landscape had appeared utterly lifeless, but it was not. There were other living beings in this bleak world.

At an altitude of 3,100 feet (945 m), they made their first camp on the plateau. Behind them, mountains dropped away into the frozen sea of the Denmark Strait. In front, ice stretched away to the far horizon. Now, Myrtle thought, it was easier to go on than to turn back. For the next few weeks, life would be reduced to the simple necessities of cooking, eating, hauling and sleeping. She wondered how many miles a day they would travel.

At this point, they transferred the loads onto one sledge, as planned, and left the second sledge behind. Starting at 3 am, in full daylight, they donned their harnesses and heaved it forward. They were off.

It wasn't long before a blast of wind struck them, blowing straight down from the north. Myrtle pulled her hood as tight as possible around her head. Her face covering, an improvised strip of fabric to protect against frostbite, froze solid and flapped against her skin. Suddenly, her feet broke through the crust of snow and she was floundering in deep powder. The tight sledge line yanked at her shoulders and jerked her neck. On the other side, Bill was in similar difficulty.

With Hugh and Roger heaving on the front of the sledge, Myrtle and Bill fought to move forward in unison. It was an exhausting and repetitive pattern: one or two steps on the surface and then they sank up to their knees. Each time, the sledge slid into their footprints and had to be dragged out. Myrtle began to despair. But Hugh's philosophy was that successful polar travel depended on the simple act of keeping going. Bill started singing the Tay Boat Song and Myrtle joined in, focusing on putting one foot in front of another.

Supper that night consisted of kidney soup, made with soup powder and packed snow, followed by stew consisting of onions, margarine, a squeeze of Marmite, more snow, and four ounces each of dried meat. A bar of chocolate each was a much-needed treat. Myrtle fell asleep listening to Hugh and Roger debating whether Peary had actually reached the North Pole in 1909.

When Fridtjof Nansen made the first crossing of the Greenland ice-cap in 1888, he wrote: 'To give a description of the scenery during the day in this region is easy. We saw only three things: that was snow, sun, and ourselves.'[2] Myrtle's experience was very similar. The sky was such an endless, unbroken blue that she got excited when a cloud formed and broke up its monotony. Endlessly sliding one ski in front of another, with the sledge lumbering behind, she daydreamed about a holiday lying in the sun, and used her imagination to design a new house.

Their average speed of one mile per hour rose to two; the distance was gradually being eaten away. At regular intervals, they stopped to take compass, sextant, altimeter and milometer readings. Roger made weather observations, but they didn't expect the rain that kept them in the tent for a whole day and a night. When it stopped, the water froze and coated everything in a slippery glaze, giving no traction on skis. They bound lengths of climbing rope underneath their feet, but they still took one step forward and slipped half a step back.

Then the wind got up again, cutting through Myrtle's layers of wool jersey, cashmere and Viyella, topped with an anorak and a nylon cagoul. With blowing mist limiting her world to just a few feet all around, she had a curious feeling that they had an extra companion - a 'fifth man', akin to the 'fourth man' sensed by Shackleton, Tom Crean and Frank

Worsley in South Georgia. She later found that Bill and Roger had unusual sensations too: they either expected to see a tree or a hedge looming out of the mist, or felt they were travelling in a deep groove with banks on either side.

On Myrtle's birthday, in early July, a box of Black Magic chocolates and a bottle of Drambuie were brought out, and chocolate ice-cream was made by stirring cocoa and dried milk powder into a bowl of snow. It was a well-deserved celebration in many ways. They were now averaging two miles per hour, and their readings showed that they had almost passed the highest point. Theoretically, it would be downhill from now on.

A dark feature on the horizon, no more than a hazy blur, eventually materialised into the US Air Base at Sondrestrom. Located at the east end of the Kangerlussuaq fjord, this was one of the Distant Early Warning Line bases set up in response to the Cold War threat. It took a while to convince the staff that these wild-looking visitors had actually walked across Greenland, but soon they were swept inside and plates of sandwiches and cream cakes, fresh oranges and steaming cups of coffee were put in front of them.

Talking to the officers, it dawned on Myrtle that most of them never went outside except to board a plane. A phone was put in her hand and she was urged to call home, but her mind went blank; Hugh remembered a number, but when Myrtle's astonished mother answered, she reminded them that it was two o'clock in the morning.

For the final stretch they tried mounting a sail on their sledge, like Nansen had done, and it sped along so fast that they were nearly bowled over. The western edge of the ice-cap was fringed with the same kind of obstacles that they'd faced at the start of their journey: crevasses and dunes of ice, giving way to partly-frozen bog. At a glacial lake, the sound of running water delighted them, until it started seeping into their boots. Myrtle took off her gloves while hauling the sledge over a fast-flowing river, and her hands were immediately frozen into a clenched position. She rammed them quickly into her gloves to thaw out.

While Hugh and Roger grappled with the sledge, Myrtle and Bill were sent off as scouts, tasked with finding dry land. On the far side of a heap of moraine, they found it. A snow bunting was singing above a carpet of purple saxifrage and moss campion. Myrtle shed happy tears as she bent to bury her face and hands in the flowers.

A few days later, three small bundles of excitement were clambering down the steps of a plane and throwing themselves into their parents' arms. For a few weeks, while Hugh, Bill and Roger extended their journey up the west coast by kayak to Christianshåb (Qasigiannguit), Myrtle camped at Angujartorfik with the children and her friend, Heather, who had travelled with them. Alongside Magdalena and Jacob, a Greenlandic couple, and their children Gertrude, Kerstinak and Pok, they fished in the rivers, cooked reindeer meat and gathered berries. The children played with seaweed, slid down sandbanks, chattered endlessly together in their respective languages, and squeezed themselves, giggling, into a single sleeping bag.

At the time, Myrtle encountered criticism in the press for taking her children to Greenland because the risks were perceived to be too great. But, she says, you can find solutions to those risks before you leave home, if you plan ahead wisely. Food, she knew, would not be a problem: 'You're not going to run out of food in west Greenland if you go at the time when the salmon are coming up the rivers.'[1]

Looking back at her achievement, Myrtle explains: 'I never thought that I'd end up as the first woman [to ski across Greenland]... and I've never really met anyone who I would call a real explorer, who does think like that. You've gone because you're passionate about the land, or the people, or what grows there.'[1] Summing up her attitude to expeditions, and to life in general, she says: 'You've got to have a North Pole to head for - all of us, at any age. You've got to be glad to be alive, and look to the future.'[1]

The Greenland summer of 1965 turned into autumn, and it was time to say regretful farewells to new friends. From a plane heading homewards, Myrtle gazed down at the Inland Ice. 'See that?' said the passenger sitting next to her. 'Some damn fools crossed it on foot this year. How silly can you get?'[3]

o

In 1958, with Hugh Simpson and Bill Wallace, Myrtle made several new ascents in the Andes. In 1969, Myrtle and Hugh, together with Roger Tufft, attempted to ski to the North Pole, but had to abandon the expedition at 84°N when their generator blew up.

o

[1] Myrtle Simpson, interviewed by Sue Stockdale for *Access to Inspiration* podcast (November 2022)

[2] Dr. Fridtjof Nansen (1889) 'Journey across the inland ice of Greenland from east to west', *Scottish Geographical Magazine* (1889)

[3] Myrtle Simpson, *White Horizons* (1967)

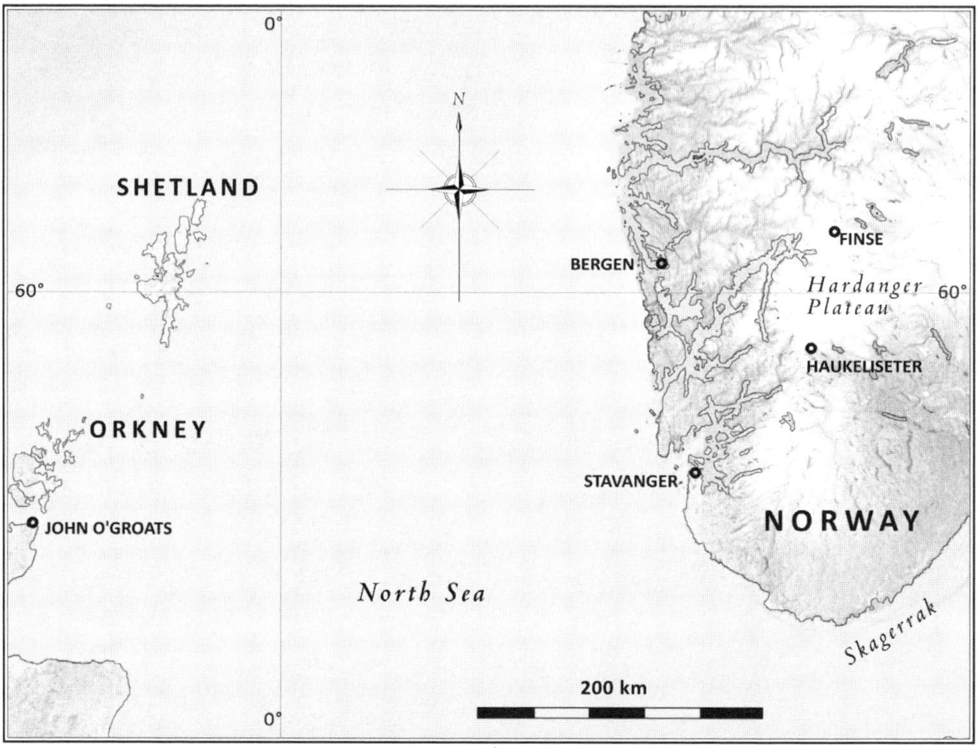
0°
N
SHETLAND
FINSE
BERGEN
60°
Hardanger
Plateau
60°
HAUKELISETER
ORKNEY
STAVANGER
JOHN O'GROATS
NORWAY
North Sea
Skagerrak
200 km
0°

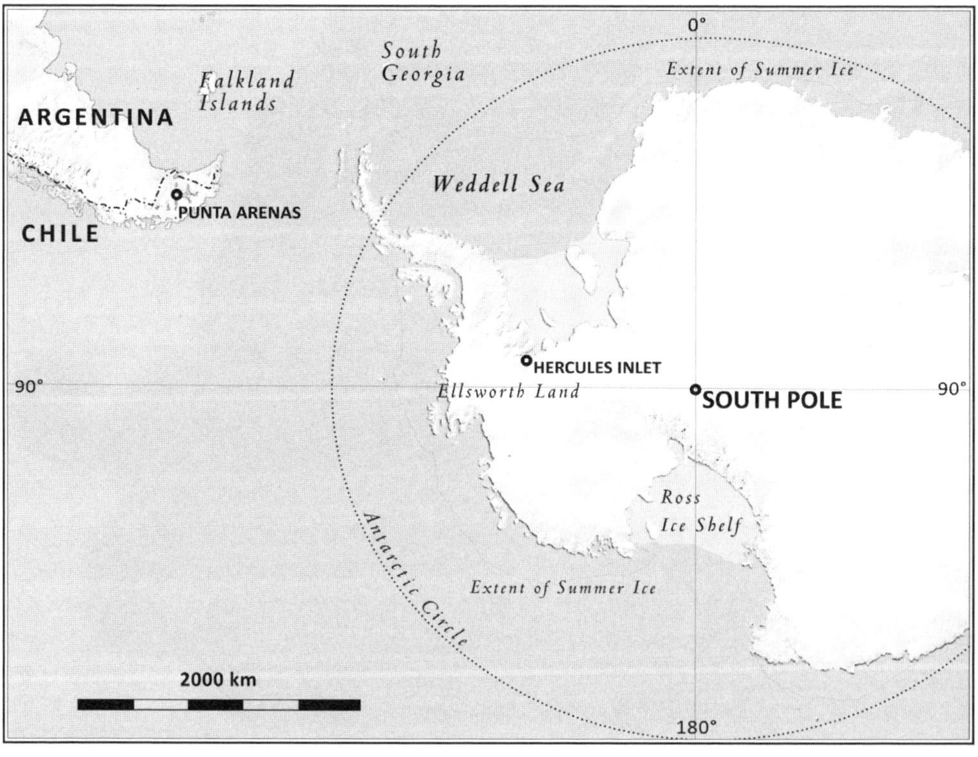
0°
South
Georgia
Extent of Summer Ice
Falkland
Islands
ARGENTINA
Weddell Sea
PUNTA ARENAS
CHILE
HERCULES INLET
90°
Ellsworth Land
SOUTH POLE
90°
Ross
Ice Shelf
Antarctic Circle
Extent of Summer Ice
2000 km
180°

Luke and Hazel Robertson

Hazel Robertson reflects on why she and her husband, Luke, are drawn to the Earth's cold, wild and remote places. 'I think it's the almost endless horizon, just the sky and the ice. It's knowing that we have everything with us that we need to survive. We're tuning in to nature and the landscape, we're feeling snow squeaking under our skis and breathing in the cold, crisp air.' [1]

Part of Hazel's childhood was spent in Alaska and Canada, camping, skiing, fishing, and exploring the forests; as a teenager, she led youngsters on wilderness expeditions in Michigan and Ontario. Back in her homeland of Scotland, she discovered the exhilaration of trail running and completed the five-day Arctic Ice Ultra in Sweden. With her husband, Luke, she became an expedition leader with the Polar Academy, taking students on life-changing journeys in the Arctic. A strong sense of adventure is the thread running through her life.

One evening in February 2020, with snow swirling around in the darkness, Hazel stepped off the train at Finse, a tiny village in south-west Norway whose railway station is 4,000 feet (1,219 m) above sea level. Donning skis and full winter gear, she embarked on two weeks of intensive polar expedition training and then headed south, pulling her fully laden pulk in the direction of the Hardanger Plateau.

Covering an area of about 2,500 square miles (6,500 square km), the Hardanger Plateau (Hardangervidda) is the largest mountain plateau in Europe. Rocky hills and treeless moorland, randomly etched with streams and lakes, are blanketed in winter by deep snowfall. Hazel intended to make a solo crossing of the plateau, a journey of just over 80 miles (130 km). Physically and mentally, she was ready.

The beginning of an expedition, explains Hazel, always brings a sense of freedom and self-reliance: 'Life gets so simple. Everything is stripped back. You have your food, you have water - the snow that's all around you, and you've got a stove so you can melt it - and you've got shelter, which is your tent. As long as you can stay safe in these environments, you can keep going. It's an amazing feeling, knowing how little we need to actually survive.' [1]

'It's an amazing feeling, knowing how little we need to actually survive.'

But a serious and unprecedented challenge was heading her way. In February 2020, three Atlantic storms in quick succession battered the shores of north-western Europe. The UK experienced its highest monthly rainfall since records began. On the Hardanger Plateau, all this precipitation fell as snow. Every morning, Hazel had to dig her tent and her pulk out of deep drifts. Every evening, she camped at the base of the next hill she had to climb. And every night, the weather forecast, received via satellite, told her only one thing: more snow.

Often it would take all her strength just to get the pulk moving, and for the rest of the daylight hours she was lugging it on a zigzag path upwards; only rarely did the clouds break to give spectacular views across the plateau. 'I had the chance,' she says, 'to get really frustrated, especially with day after day of snow... but I remember thinking, that's not going to get me up the hills any quicker.' [1] So she broke her journey right down, counting 10 steps, then another 10, and then another, stopping briefly each time to rest and celebrate. 'Four or five hours later,' she says, 'I'd be at the top of the hill, and I'd done it 10 steps at a time.' [1]

Hazel completed her journey from Finse to Haukeliseter in 12 days. It took longer than anticipated, but the lesson she learned was far more important. 'There are so many things in life outside our control,' she says. 'Don't waste energy on things you can't change. Focus instead on what you *can* control: how you think about the situation you're in.' [2]

In rain-soaked Scotland, Luke had been watching Hazel's progress in the form of GPS dots on a map. Four years previously, their roles had been reversed.

Growing up on a farm in rural Aberdeenshire and school trips to Iceland and Switzerland with his geography class had given Luke a passion for the outdoors. Stories about polar explorers of the Heroic Age - Nansen, Scott, Shackleton - made him curious about the Antarctic. Although he struggled to imagine the reality of it, he thought how much he would love to go there, some day, on a big adventure.

But then fate took a hand. Aged 23, he was diagnosed with a heart condition, and to save his life an artificial pacemaker was fitted. Shock and fear were his first reactions. Then came another realisation. 'Life,' he says, 'is about a series of moments and the choices we make in those moments are what determines our future.' [1] Luke chose to focus on what he wanted to do with his life, and took a few steps closer to his Antarctic adventure.

Just before heading off to Norway and Greenland for expedition training, Luke was told that he had a brain tumour. Once again, an operation was essential.

Lying in his hospital bed, he could see the sun setting behind the Pentland Hills, and imagined the same sun guiding him to the South Pole. He woke up from the operation to be told that the tumour was, in fact, a benign enterogenous cyst. Discharge from hospital came sooner than he'd dared to hope, and he discovered that his ambition now had a deeper purpose: 'I wanted to inspire other people who were going through their own difficulties to realise that they were capable of so much more - to come through those difficult times and do things that they, at some point, only dreamed could happen.' [1]

On 5th December 2015, Luke stood on the shore of Hercules Inlet in Antarctica. The plane that had dropped him off disappeared into the clouds, and the silence of the ice enveloped him. For many months, he and Hazel had planned every last detail of this expedition. In Punta Arenas, they had spent days decanting food into zip-lock bags, daily rations that were calculated down to the last gram. Clothing, fuel, food, equipment and shelter had been loaded onto a sledge that now weighed 130 kilos.

Now here he was, on the White Continent. This was the vision that he had been determined to make into reality, the dream that had given him hope. He was proposing to ski alone and unsupported to the South Pole, which was 730 miles (1,175 km) away.

Anxious questions jostled together in his mind. What if his tracker failed? What if his ski bindings broke? Had he packed enough fuel? Would the canisters leak? His charity fundraising campaign had already reached £35,000, and he wondered if he was worthy of all that faith in his ability to succeed. As he hauled on his sledge and took his first steps southwards, he consciously slowed his thoughts and remembered his mantra: *everything simple and deliberate.*

For the first few days, the weather was appalling. Fierce winds battered his tent, and the energy expended from pulling the heavy sledge meant he was breaking into food rations that were intended for much later in the journey. In them were secreted little notes written by Hazel, telling him that he hadn't got far to go now, although he was still hundreds of miles away. And then, two weeks into the expedition, Hazel got an unexpected call. Luke's solar panel had failed. He was relying on it to charge the batteries for all his gear, including his satellite phone and tracking device. From now on, the charge left in those batteries would have to last until the end of the journey.

The Antarctic Logistics team offered Luke a 'tech drop' - access to a buried cache of satellite phones and batteries at the half-way point. Luke considered it, and declined. His aim was to ski to the Pole unsupported, which meant no outside help or re-supply. He says: 'It was quite a big decision, but I stuck to it. When I skied past that tech drop, I remember thinking, this is now or never.'[1]

It was ironic, Luke reflected, that the exploits of those early polar explorers had fascinated him so much. He had wanted to follow in their footsteps, but he hadn't expected to endure a similar lack of communication with the outside world. Regular phone calls with Hazel had been allowing her to share updates on social media, but now all she knew was his nightly position, supplied by the logistics team who were monitoring his tracker. Conserving battery life was paramount, because if his position wasn't logged for two consecutive nights, a plane would be sent out to look for him and the expedition would be over. The early explorers had never had that kind of back-up.

Again, worries invaded Luke's mind. Could he really last another 20 or 25 days without charging equipment? What if he was taken ill? Quickly, he shifted his focus. He could, at least, have a positive mindset. He could eat a good diet and focus on staying healthy. Small rewards made an enormous difference: 'On Christmas Day, I gave myself a little boost and changed into my only clean pair of underwear. One of the best Christmases ever!'[1]

Battling strong headwinds and temperatures that reached -50°C (-58°F) with windchill, Luke made the slow and gradual climb up onto the Antarctic Plateau, which has an average elevation of 9,800 feet (3,000 m). With only about a hundred miles to go, he was running short of food and his body was dangerously low on energy. One night, he dreamed he was pushing a shopping trolley around a supermarket, and filling it with everything he craved. Then, in the waking hours, he started to see and hear things: animals and cartoon characters running around him in the snow. Sometimes he heard his name being called from behind, and couldn't resist glancing back, because, as he explains, 'If I hadn't looked, and someone had tapped me on the shoulder, that would have been the scariest thing imaginable!'[1]

One night, he dreamed he was pushing a shopping trolley around a supermarket, and filling it with everything he craved.

In reality, he was getting very little sleep, and at the same time he was doubling or tripling his daily mileages. Later, a Professor of Psychology, Ben Alderson-Day, explained to him that the mind will sometimes create what it needs in times of stress. Luke certainly needed more food, and he was seemingly missing the company of people and animals, and the sense of comfort they bring.

But on 13th January, one sight that met his eyes was definitely real and solid. In the distance were the buildings of the Amundsen-Scott South Pole Station; in front of them, amid 12 national flags representing signatories of the Antarctic Treaty, stood a short ceremonial pole. Luke's gloved hands rested on it for a few seconds and then he raised his arms in relief. He had made it.

The South Pole expedition, says Luke, taught him 'so much about the power of exploration - not just by raising funds for amazing charities, but also the power to bring people together behind this cause.' [1] From all over the world, he received messages from people who were overcoming their own health issues and had been inspired by his story. For Luke, this made everything worthwhile.

In the spring of 2017, Luke and Hazel set off on a joint expedition with the aim of travelling by human power from the southernmost to the northernmost point of mainland Alaska - a distance of over 2,000 miles (3,218 km), beginning in temperate rainforest and passing glaciers and mountains before reaching the Arctic tundra. It was an experience that forced them to abandon their expectations and reconsider the landscape that they were passing through.

Luke explains: 'We'd gone there looking to understand how climate change was affecting people, and all through the journey we were seeing examples of it.' [3] In the Yukon, the outflow from a dwindling glacier had caused a river to change its course, so that, almost overnight, local people who relied on fishing for sustenance found themselves with no river to fish in. Rising sea levels and storm surges are eroding the coastline so quickly that even the latest maps bear no relation to the terrain. Entire villages are being forced to re-locate.

Following an ancient Inupiat trading route, Luke and Hazel had planned to navigate a network of lakes and connecting streams in the far north. But they found that thawing permafrost had drained them, so instead they waded through mud, hauling or portaging 200 kilos (440 lb) of gear. At a lake that was filled with waist-deep sludge they set up camp in silence, knowing that they could not continue. Reflecting on the immediate feeling of failure from that experience, Luke says: 'I came to realise that exploration is really about discovery, not necessarily about completing what you set out to do.' [4]

On Anachlik Island in the Arctic Ocean, Luke and Hazel stayed with the Helmericks family who still live largely off the land, using their intimate knowledge of nature, the weather and the seasons to

Moose Factory, HBC base where John Rae was stationed. *Illustration by William Trask (1854)*

Hall of Clestrain, John Rae's family home in Orkney
Colin Woolf

'Theodolite work in the pack', illustration from C E Borchgrevink's *First on the Antarctic Continent (1901)*

The stern of Shackleton's ship *Endurance*, photographed by the Endurance22 Expedition
Falklands Maritime Heritage Trust and National Geographic

'Trying to cut a way for the ship through the ice to a lead ahead. Feb 14, 1915.' *Endurance and crew, by Frank Hurley from 'South' by Sir Ernest Shackleton*

Hugh and Myrtle Simpson and Roger Tufft in Greenland
Hugh Simpson

Hazel Robertson's camp on the Hardanger Plateau
Hazel Robertson

Luke Robertson at the South Pole
Luke Robertson

Col John Blashford-Snell near Shafartak Bridge at the end of the Blue Nile Expedition
Chris Bonington

Tisisat Falls, photographed on Col John Blashford-Snell's Blue Nile Expedition
Scientific Exploration Society

Benedict Allen's first encounter with the Yaifo, Papua New Guinea

Benedict Allen

Sal Montgomery paddling the Río Claro in Chile

Casey Bryant Jones

The Glenfeshie Estate
Alex Baxter

Ray Mears demonstrating bushcraft
Woodlore

Forest fire in a gorge near Tempathang, taken on the Scottish Women's Himalayan Expedition *Monica Jackson, courtesy of Simon Jackson*

Monica Jackson and a Sherpa companion on the Scottish Women's Himalayan Expedition
Monica Jackson, courtesy of Simon Jackson

survive. Afterwards, they wrote: 'Dependency on this wild environment… brings with it a requisite for respect. Perhaps we can all learn a thing or two from this family, by seeking to understand where our resources come from and where they go, and by taking only what we need.'[5]

Now at home in Scotland, Luke and Hazel continue to encourage others to get outdoors and embrace challenges with a positive mindset, while appreciating the value of skills, confidence and knowledge acquired along the way. Hazel says: 'The outdoors is perhaps the greatest of life's teachers. It was where I learned to fully trust myself. The simplicity of life on expedition soaks into my bones and fills me with happiness.'[2]

1 Luke and Hazel Robertson, 'Reflections on Ice', online talk for RSGS (November 2021)

2 Hazel Robertson, 'Six things the outdoors has taught me about life', LinkedIn post (September 2021)

3 Luke and Hazel Robertson, interview with Jo Woolf for RSGS (October 2021)

4 Luke and Hazel Robertson, Due North: Alaska Expedition Report for RSGS (2017)

5 Luke and Hazel Robertson, 'Lessons learned from a lifetime in the Arctic', blog post on everyday-exploring.com (December 2019)

FOREST AND RIVER

Among the scenes which are deeply impressed on my mind, none exceed in sublimity the primeval forests undefaced by the hand of man… No one can stand in these solitudes unmoved, and not feel that there is more in man than the mere breath of his body.

Charles Darwin, *Narrative of the Surveying Voyages of His Majesty's Ships Adventure and Beagle between the years 1826 and 1836*

Ah me! Give me a West African river and a canoe for sheer good pleasure.

Mary Kingsley, *Travels in West Africa*

SYRIA
Mediterranean Sea
LEBANON
IRAQ
ISRAEL
JORDAN
KUWAIT
LIBYA
EGYPT
SAUDI ARABIA
Red Sea
The Nile
CHAD
SUDAN
Khartoum
ERITREA
YEMEN
DJIBOUTI
SOMALILAND
CENTRAL AFRICAN REPUBLIC
SOUTH SUDAN
ETHIOPIA
DEMOCRATIC REPUBLIC OF THE CONGO
UGANDA
KENYA
SOMALIA
TANZANIA
Indian Ocean
1000 km
35°
The Nile
SUDAN
KHARTOUM
Lake Tana
Nahr ad Dindar
12°
12°
BAHAR DAR
Tis Issat Falls
Blue Nile
White Nile
DEBRA MARK'OS
Mt. Talo 3480m
Northern Gorge
Western Cataracts
Blue Nile
Shafartak Road Bridge
ADDIS ABABA
ETHIOPIA
N
200 km
35°

Colonel John Blashford-Snell

o

Colonel John Blashford-Snell has been described as one of the great explorers of our age, but he very nearly didn't live long enough to do any exploring at all. At their school on the island of Jersey, he and his friends longed to dive underwater like their hero, Jacques Cousteau. A self-contained breathing apparatus known as an Aqua-Lung had been invented by Cousteau and Emile Gagnan in 1943, but such sophisticated equipment was out of reach to John and his friends, so they decided to make one themselves.

In Jersey, a lot of military equipment had been abandoned after the German occupation during the Second World War. One of John's friends got hold of a Dräger kit, which was a submarine escape apparatus with a canister of soda lime for turning carbon dioxide into oxygen, and he converted it for diving underwater. John was tasked with trying it out. He recalls that his friend 'put this thing on me, rather like a gas mask, and tied a belt round my waist with some nets full of house bricks to act as weights, and sank me in the sea just off St Brelade's harbour.' [1]

After a short while, breathing became almost impossible: they had substituted caustic soda for soda lime, and the resulting gas was awful. John tugged hard on the rope that was attached to him, as a signal to be pulled back up, but was horrified to see the other end of the rope sinking slowly down to the sea bed. He tried, and failed, to offload the nets of house bricks, and was forced to make an underwater climb up the harbour wall, while struggling to breathe. When he emerged, gasping, he saw his friend chatting to two girls at the top. 'He'd completely forgotten about me!' [1]

For John, adventure was something of a family tradition: his parents had led groups of scouts and guides into the Blue Mountains of New Zealand, and his godfather's proud boast was that he had ridden a horse from Cape Town to Cairo. His school actively encouraged outdoor pursuits such as climbing and boating. Despite the unfortunate incident with the 'Aqua-Lung' John entered the Royal Engineers and was initially posted to Cyprus, where he became an Army diver.

John's godfather had a strong link with Haile Selassie, the Emperor of Ethiopia. Consequently, John found himself invited out to Ethiopia, in charge of Sandhurst officer cadets on natural history expeditions.

Shortly afterwards, a letter from the Emperor saying he wanted his 'Great Abbai', the Blue Nile, explored, arrived at the Royal Military College of Science where John was a student. He was summoned by his General. Had he seen the Blue Nile? What did he think of it? John replied that he had only seen it briefly, but it ran through gorges a mile deep, was infested with crocodiles and hippos, and was known to be patrolled by bandits. This turned out to be just the answer that his General was looking for, and the Blue Nile Expedition was born.

Even in 1968, large stretches of the Blue Nile were still unmapped. Since the 1700s, a succession of would-be navigators and surveyors had found that raging through its steep-sided gorges were the kind of rapids that no boat could possibly survive. That was the accepted belief - but by this time, John was already pondering a possible solution. As he stood and watched the Blue Nile tumbling 150 feet into a canyon at the Falls of Tisisat, it occurred to him that a descent of the worst rapids might be possible if they used inflatable boats.

'Don't go near hippopotamus in the mating season,' was the advice from scientists at the Natural History Museum. That was all very well, but did they know when the mating season was? They weren't sure, but they were hoping the Blue Nile Expedition would shed some light on it. Meanwhile, John heard horror stories about crocodiles of gargantuan size, venomous snakes so plentiful that he would need long leather boots, water-borne parasites inflicting serious diseases, and even landslides leaking radioactive gas. Undeterred, he selected a team of military servicemen and scientists, and they flew to Addis Ababa in July 1968.

'Don't go near hippopotamus in the mating season.'

The upper stretches of the river were accomplished, as far as possible, in inflatable boats, backed up by larger Army assault craft. Where the rapids became too severe, the boats were portaged around the sides - a hazardous task in itself - and they were lowered, empty, down the Tisisat Falls by rope. Tragedy struck one of the boat parties early on: attempting to put a rope across a tributary that was swollen by recent rain, Corporal Ian Macleod was swept away. Desperate efforts to reach him proved futile, and a search downstream by helicopter found no evidence. In deep shock, the team gathered themselves together and decided to carry on.

In the Northern Gorges, the party encountered their first real threat from human presence. In a creek, where he had gone ashore with a geologist, John heard a bullet smack into the rocks above his head. A man with a rifle was standing on the clifftop above. John recalls: 'I shouted at him, "Watch out, you might have hit me!" And I suddenly realised, that's what he was trying to do!' [1]

After that, a hostile situation developed quite rapidly. John had been prepared for confrontations with armed anti-government rebels, but when a group of about 40 men appeared and opened fire, he judged it best to take to the boats and get everyone safely downstream. That night, their camp on an island was attacked, forcing John and his team into defensive action followed by a strategic retreat. Negotiating rapids in pitch darkness was not a scenario he had hoped to face. One of the boats flipped over, and John's own boat, having sprung a leak, was sinking fast. The helmsman headed hastily for shore and crashed into a sandbank. John was tipped out, and was startled to land on something long and scaly. An equally startled crocodile shot off into the water.

Reunited on the sandbank before daybreak, the men made some makeshift repairs to their vessels and re-launched themselves into the flow. The river widened into a quieter stretch, and early sunlight glanced off the mirror-like surface. It was calm enough, in fact, to allow good views of a whole regiment of crocodiles, some about 18 feet long, pushing off like torpedoes from the bank and heading in their direction. John grabbed a rifle, but Colin Chapman, one of the zoologists, assured him that they were merely curious. The crews held their breath, listening to the sound of scaly backs scraping against the bottom of the boats, but Colin was right: eventually, the crocodiles drifted away.

The Blue Nile, however, had one more surprise to throw in their path. John explains: 'We came round a corner, and the whole river was full of hippopotamus!' [1] Acutely aware that their black, rubber-hulled boats bore a worrying resemblance to female hippos, the team edged cautiously past the lurking monsters. When *was* the hippo mating season, exactly? Were they about to find out? Luckily, the time didn't seem to be right. For whatever reason, the hippos merely grunted and watched them go.

As John began to compile his report for the Emperor of Ethiopia, and his scientists set to work on academic papers about their discoveries, it occurred to him that there was a need for a permanent organisation to mount further expeditions of this kind. This is how the Scientific Exploration Society was born. John explains: 'The fact was that the services had a lot of hell-bent adventurous characters who were good at field administration and could handle logistics. We'd already established a very good relationship with the Natural History Museum and also with people like Kew Gardens who were eager to get out [to such places]. So the idea of the SES was to bring together the service explorers and the scientists to work together for the general benefit of mankind.' [1] The SES continues to initiate a worldwide programme of scientific expeditions in remote parts of the globe, with a focus on research, conservation, education and community aid; it is one of the longest-running exploration organisations in the world.

Film footage of the Blue Nile Expedition was made into a television documentary, and as a result, whitewater rafting became a popular outdoor pursuit. John remains steadfastly modest, admitting only that he invented whitewater rafting 'by accident.' He adds: 'I've got one of [the inflatables] in the workshop here that goes back to 1968, and it still floats!' [1]

John has always been curious about tales of mysterious creatures surviving in obscure and isolated regions. For example, in the Terai region of southern Nepal, local people reported an elephant-like monster with a huge domed head and extraordinarily long tusks that was believed to be a living mammoth. John took teams of scientists to investigate, and their conclusion was that, although this wasn't a mammoth, it was one of the largest Asian elephants ever studied, one of a herd of about 50 such animals that were living in the forest.

And when he led an expedition in Bolivia, John was carrying with him a book by Colonel Percy Fawcett, who claimed to have discovered the Double-nosed Andean Tiger Hound there in 1913. Back in Britain, Fawcett's story about a dog with two nostrils was greeted with derision, but John was keeping an open mind. One evening, he was chatting to a local schoolmaster while a group of dogs lounged close by. Something about the face of one of them struck John as odd, and he went to have a look. It had divided nostrils. The schoolmaster revealed that he'd bought the dog, and that there were similar animals locally. John was astounded: 'I said, "What do you use it for?" "Oh," he said, "hunting - they've got a wonderful sense of smell!"' [1]

As the Scientific Exploration Society continued to evolve, expedition teams included doctors and dentists who were welcomed by communities desperate for medical aid. One such expedition focused on the Wai Wai tribe in Guyana, but at the end of the expedition one of their church priests approached John with an additional request: 'He said to me, "When you come back, as I hope you will, would you do us a favour?" And I said, "What is it?" And he said, "Would you bring me a grand piano?"' [1]

The young people of the tribe, according to the priest, had great musical talent but they were getting lured away from the rainforest by the 'bright lights' of coastal towns. A grand piano, he felt, would refresh their sense of pride and identity. Even John thought it was a crazy idea, but he promised to see what he could do. Back in London, wrapping up a fund-raising presentation in a West End hotel, he joked that the Wai Wai were looking for a grand piano. The response was immediate: not only was he offered an instrument, but an airline pledged to fly it out for him.

Getting a grand piano through the interior of Guyana was never going to be an easy job. John called on his Royal Engineer colleagues, and with local help they heaved the 800-lb instrument, which had been partly dismantled and packed in a large crate, through dense rainforest, over specially-built bridges, and even up a stretch of rapids on the River Essequibo by means of a motorised canoe. In the Wai Wai's hilltop settlement, it was unpacked and re-assembled amid much excitement, and a piano teacher immediately set to work.

John reflects that his so-called 'Grand Adventure' had even wider benefits, because the American charity Conservation International saw the film that was made of it, and took steps to protect that area of rainforest from exploitation by logging and mining. The Wai Wai became the official protectors of Guyana's first community-owned conservation area, and the young people were offered posts as rangers. 'And then,' adds John, 'they sent me another message, saying that the piano needed retuning, so I had to do another expedition, with adventurous piano tuners!' [1]

In 1978, supported by the Prince of Wales (now King Charles III), John set up Operation Drake, a two-year round-the-world voyage that allowed young people to participate in a succession of scientific and community projects. This was followed in 1984 by Operation Raleigh (later Raleigh International), a more ambitious scheme which carried young volunteers on similar expeditions all over the world. John says: 'I look back on my life so far as being a catalyst for young people, not only from Britain but all over the world, to go and do something… it's giving young people the opportunity.' [1]

'I look back on my life so far as being a catalyst for young people, not only from Britain but all over the world, to go and do something.'

Many participants in these schemes went on to achieve great things. John proudly recalls congratulating a team of kayakers in Alaska who had completed a difficult journey: 'I said, "Now you've done all this, you can reach for the stars!" And one of them did - that was the astronaut, Tim Peake!' [1] Other alumni include Dame Barbara Woodward, now the UK's Permanent Representative to the United Nations in New York, and RSGS's own Chief Executive, Mike Robinson. John also remembers a young man calling in to his office and telling him that he wanted to go up the White Nile by hovercraft: 'Anyway, off he went and did it. His name was Ranulph Fiennes, and he's been doing expeditions ever since!' [1]

In his book, *From Utmost East to Utmost West*, John also makes the point that expeditions are not just for the young. Older people, he says, have the kind of life experience and wisdom that could benefit conservation projects and assist developing nations. A candidate who is reasonably fit could have a rewarding and enjoyable experience by joining a project that is making a real difference to people in need. By continuing to take part in worldwide expeditions himself, John is a living embodiment of that belief.

◦

> The Scientific Exploration Society has initiated over 200 expeditions since its inception. It continues to promote and support expeditions focused on scientific exploration, while collaborating with under-resourced communities around the globe and seeking sustainable solutions to the challenges they face. In addition to advancing the frontiers of knowledge, its mission is to advocate for the preservation of biodiversity and indigenous cultures, and to serve as a catalyst for positive change.

◦

[1] Colonel John Blashford-Snell, interviewed by Dr Vanessa Collingridge for RSGS Inspiring People talks (2022)

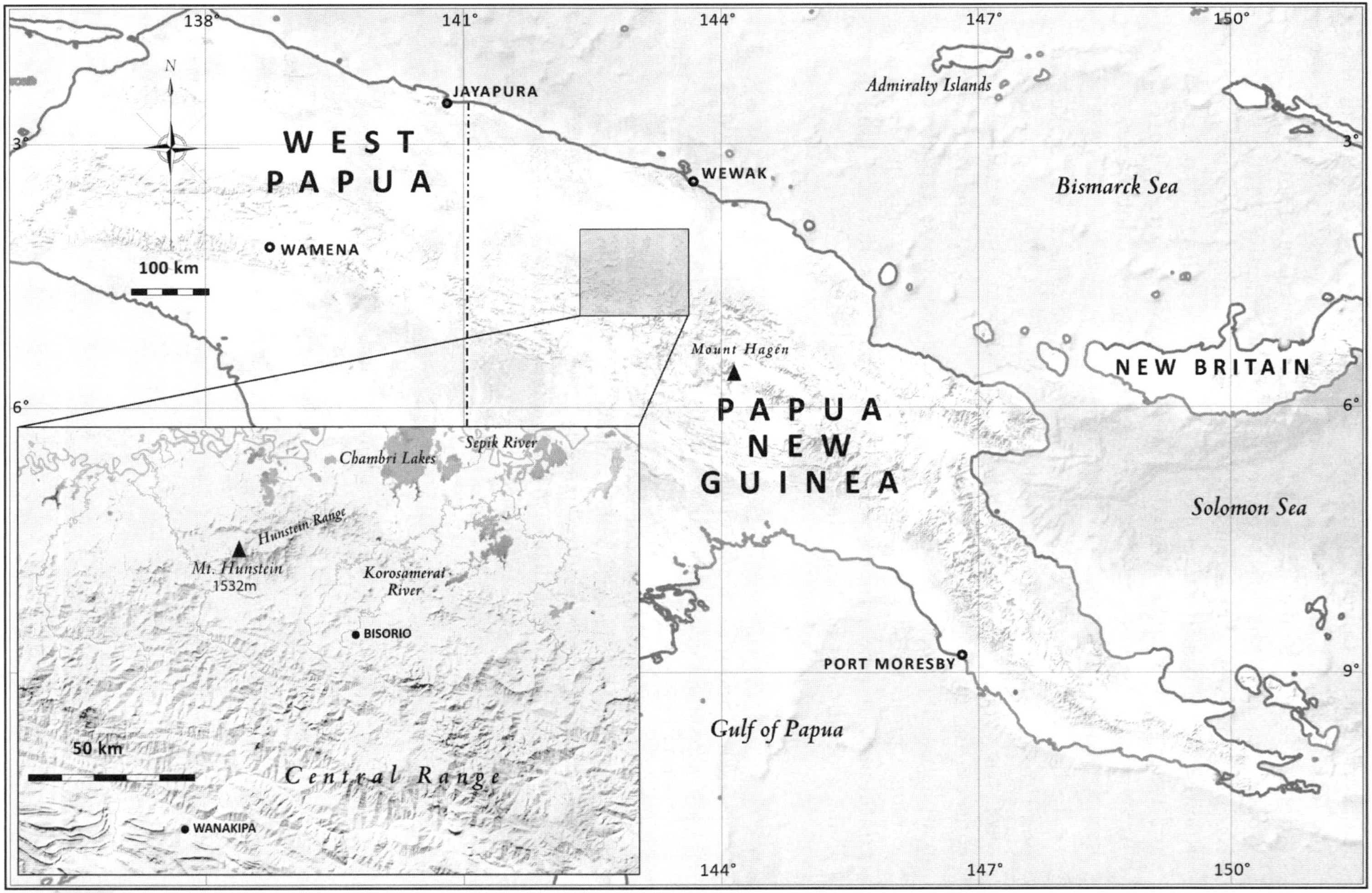

138°
141°
144°
147°
150°
N
JAYAPURA
WEST
PAPUA
WEWAK
WAMENA
100 km
Admiralty Islands
Bismarck Sea
Mount Hagen
NEW BRITAIN
PAPUA
NEW
GUINEA
Solomon Sea
PORT MORESBY
Gulf of Papua
3°
6°
9°
Sepik River
Chambri Lakes
Hunstein Range
Mt. Hunstein
1532m
Korosamerat
River
BISORIO
50 km
Central Range
WANAKIPA

Benedict Allen

On a forested mountain, high above the Korosamerai River, Benedict Allen was watching the Yaifo people dance. To the rhythmic beats of a drum, men stamped and whirled in a circle around the fire. Together, they chanted the names of natural entities that shared their rainforest world: the shy, tree-dwelling cuscus, the fruit bat, the flowing water of a stream. Brilliant plumes of feathers, jutting from their armbands and headdresses, fluttered and swayed in the firelight.

Benedict was, to his knowledge, the first outsider to witness the Yaifo's ceremonies, or indeed any aspect of their existence. And as he watched this dance, which seemed to be, above all else, a joyful and emphatic expression of their identity, he knew that he would very likely be the last.

According to the lore of the Yaifo, this rainforest had provided refuge for two of their ancestors who escaped a great flood by transforming themselves into mosquitoes and clinging to the fur of a rodent that ran for safety to the hills. Since then, they had dwelt for countless generations on the high slopes of Papua New Guinea's Central Range.

At first, when they had seen the giant birds hovering noisily over their home, the men had cleared trees to make nest sites for them. But now they understood that these were not birds at all. They were helicopters, and what they sought was gold. Alarmed and confused, the Yaifo were now preparing to leave. But before they left, they wanted to dance.

Just days before, Benedict had taken his life in his hands by trekking up the mountain in advance of the gold miners, in order to meet the Yaifo people. These were the last of the indigenous peoples to safeguard their traditions and lands, all the other groups having long since heeded the call of American missionaries and headed down to the swamplands where there was a mission station.

Benedict wanted to record the Yaifo's world before it was gone forever.

Despite knowing that they might treat him as an invader, Benedict wanted to record the Yaifo's world before it was gone forever. Now he was in their midst, and he watched and listened intently, almost choking on the thick smoke as he tried to absorb every detail. 'Remember this,' he told himself. 'If you survive to have children, tell them about this.' [1]

Benedict's curiosity and restless yearning for adventure had begun when he was a child himself. His father, Colin, was a test pilot who'd put the Vulcan bomber through its paces; watching him fly back and forth over their family home in Cheshire, Benedict began to dream of his own daring exploits, and soon he was reading tales of pioneers who had headed off towards distant horizons: Freya Stark, Fridtjof Nansen, Mary Kingsley and Thor Heyerdahl.

All through school, Benedict clung on to his dream of becoming an explorer. Then, while studying environmental science at the University of East Anglia, he was able to join research trips to Brunei and Costa Rica, and led a third trip, to Iceland. These experiences convinced him that he could undertake a successful expedition, but it transpired that he needed more than just confidence. 'The key to modern exploration,' he later explained, 'is sponsorship... The question was, how was I going to carry on after university?' [2]

Aged 22, while studying for a Masters degree in ecology at Aberdeen, Benedict hatched a plan. He would travel to South America, and cross the unsurveyed forests between the mouths of the Orinoco and Amazon rivers. Unable to secure backers, he would simply head off alone, having worked in a local warehouse to pay his airfare. He recalls: 'I had that naiveté that comes from being young - that belief that you're somehow immortal.' [3]

Crucially, Benedict believed that the indigenous people might help; they, of course, held all the knowledge needed to thrive in this seemingly hostile environment. His mother was not so sure. 'Now, you do have your survival kit?' she asked him, as he left for the airport. [1]

Initially, Benedict's solo expedition went well: almost too well, perhaps. As a preliminary test, he climbed Mount Roraima, one of South America's extraordinary *tepuis* or flat-topped mountains, which soars to 9,222 feet (2,810 m) above the surrounding forests and grassland. He had no guide, and saved himself from getting lost by the simple expedient of stringing a cotton thread along his route.

Then Benedict set out through northern Brazil, seeking the assistance of frontiersmen, missionaries and indigenous people as he went along. As he'd hoped, they did what they could to assist him, in effect keeping him alive as he made his way deeper into an environment that felt totally alien.

However, bands of illegal loggers and prospectors were moving into the region; just when his journey was almost safely completed, he awoke one night and overheard two drunken gold miners discussing how they were going to kill him. As quietly as possible, he grabbed his hammock and mosquito net, dragged his canoe out into the Iratapuru River, and paddled away into the darkness.

Three weeks later, when Benedict emerged into a cassava field on the edge of the rainforest, he was suffering from two strains of malaria and nearly dying from starvation. He had completed a journey of 437 miles across the north-east Amazon basin; it was an extraordinary achievement, but he found it impossible to feel any pride. Back at home, the question haunted him: why had he, a lone 23-year old who knew next to nothing, been allowed to survive? It was as if the forest had spared him for something else.

'I think,' says Benedict, looking back, 'I needed to psychologically come to terms with this forest that had almost killed me... I needed to understand why I was still alive - and only through the local people who obviously knew their forest in a way that I didn't.' [2] So when he was able, he returned to the rainforest: not, this time, to the Amazon, but to West Papua in Indonesia.

'I think I needed to psychologically come to terms with this forest that had almost killed me.'

Benedict's resulting encounter with the Obini people was comparatively short, but remains the only recorded contact between these remote forest-dwellers and the outside world. The men, he recalls, carried bows and arrows and were always on edge: not only were they living in a state of near-constant warfare with their neighbours, but they were conscious of the missionaries, miners, loggers and government officials who surrounded them and were about to change their way of life forever.

On the other side of the border, in Papua New Guinea, Benedict met the Niowra people and stayed for much longer. Central to the Noiwra's traditions was a belief in an ancient crocodile-ancestor, known as Avookwaark; Benedict eventually joined other young male initiates in a secret scarification ritual that marked their shoulders and backs with hundreds of stipples resembling the skin of a crocodile. Then came six weeks of isolation, during which he and the other young men were beaten five times a day. 'By the end,' he says, 'we knew ourselves - we knew our strengths and weaknesses so well.' [2]

In the eyes of the Niowra, Benedict now belonged there: according to the elders, this six-foot-four giant with pale skin and fair hair had come from the land of the spirits to be one of them. It was assumed that he would marry a local girl; his honorary father, Jonny Gowi or 'Eagle', reminded him that the Niowra have no word for goodbye, because no one ever leaves. But while it was a huge privilege to be accepted there, Benedict knew that he couldn't stay. This wasn't his culture, and it could never be his true home.

It was in 1987, while exploring the lowlands of Papua New Guinea, that Benedict first heard of the Yaifo people. Missionaries based in Bisorio at the foot of the Central Range told him that the Yaifo were the only highland community who had refused to come down from the mountain, and now it was too late. Gold had been found: helicopters were surveying the area, and the mission station was at a standstill as adults and children dug around feverishly, hoping to discover untold wealth.

Soon this gold rush would extend upwards, into the cloud-drenched forests, where no outsider had yet been. Benedict reflected that, for the Yaifo, the Anthropocene era was only just about to arrive, and it would have a devastating impact. Theirs was possibly the most isolated settlement in the whole of Southeast Asia, and now it would simply disappear. Benedict and an indigenous friend, Tsogomoi, decided to try and record what they could of the Yaifo before their way of life was lost forever.

Presently, news arrived that two young Yaifo men had been seen scouting around a new mining site, no doubt trying to understand what was going on. Quickly, Benedict and Tsogomoi sent word that they would like to go with them to their village, and to Benedict's surprise, they agreed.

Recruiting Michael, a young kitchen hand at the mission station, to help carry bags, the little party struck out into a region that was still almost uncharted. Overhead, tree kangaroos and cuscus gazed down as they struggled through thick forest in the sweltering heat. Occasionally they heard the calls of hornbills and birds of paradise, and saw flashes of brilliant plumage among the branches. After four or five days they smelled wood smoke, and the two Yaifo guides put down their packs. They had arrived.

At first, there was only silence. The guides called out several times, and then listened. Quite suddenly, out of the trees, came an explosion of sound and movement. A party of about 25 Yaifo youths dressed in skirts of leaves, their bodies painted black and red, began leaping and whirling around the strangers, emitting threatening yells and thwacking their bows as they readied them to fire. A young boy among them snatched a bush knife from Tsogomoi and stabbed the ground with it repeatedly, giving vent to his people's anger.

This, thought Benedict, could end badly. He stood motionless and waited. Somehow, they would have to explain themselves, if they weren't killed first.

Gradually, the tumult died away. The warriors regarded them in expectant silence. Benedict stepped forward, nodding in what he hoped what a friendly manner towards the nearest men. Unsure what else to do, he offered his hand to one, and inspected the arrows of another. Tsogomoi plucked up the courage to explain that they were merely visitors from the outside world, who wanted to record the Yaifo's lives for posterity.

At first, the Yaifo were suspicious: what made their lives so interesting, that they should attract this attention? They were perplexed, too, when they heard that the tall white man wished to find a route up and over the windswept mountain ridge, a request that they dismissed as impossible.

But after a while, the tension eased and the Yaifo women came out of hiding. Curiosity took the place of fear. Soon, everyone was examining Benedict's clothing, touching his white skin and marvelling at his lemon-scented soap, which was thought at first to be his food. Later, when Benedict wrote in his journal, a boy named Feefee ran his fingers wonderingly over the smooth paper, tracing the impression of the words.

Benedict was observing, too. Over the next few days he noted all he could: the names of those present, and the structure of their single wooden building, raised above the ground and accessed by a notched pole that could be kicked away to thwart enemy attack. He saw the men sharpening their arrows with a rat's tooth, and at night he watched them prepare for sleep by tucking away the decorative parrots' feathers that they wore in their nostrils. Every morning, the women gently combed through the ashes of the previous night's fire, looking for a glowing ember with which to kindle a new one.

A few days passed. One evening, Benedict understood that a decision had been reached. The Yaifo would leave the next morning. He sat and watched them dance around the fire, celebrating their life on the mountain for one last time.

But on the morning of departure, a man called Korsai stepped forward and picked up Benedict's rucksack. Korsai was a quiet, gentle character who had spent time sitting with Benedict and teaching him some of what he knew. Now, surprisingly, he headed uphill instead of down. It dawned on Benedict that this brave man, and another called Yado, were going to lead him over the ridge, somewhere they'd never been before, and down into safety on the other side. It was an act of extraordinary kindness to a stranger, at a time when their own future looked unavoidably bleak.

For four days, the little group braved increasingly cold conditions, scaling steep mossy cliffs and snatching what sleep they could in caves. All the while, Korsai and Yado were dressed only in the leaves that they wore around their waist. On the far side of the ridge was a settlement of the Hewa people, and it was here that Benedict's Yaifo companions bade him farewell. He stood and watched them climb back up towards the clouds, stepping lightly as they went. He doubted he would ever see them again.

Through the years that followed, Benedict undertook many other explorations with the help of indigenous people. He heard no more of the Yaifo, but his deep sense of gratitude and indebtedness to them remained.

About 25 years later, Benedict found himself back in Papua New Guinea, helping BBC correspondent Frank Gardner fulfil a lifelong dream to see birds of paradise in the wild. The missionaries of Bisorio were long gone, but Benedict bumped into Michael, his original guide. To his utter astonishment, he learned that the Yaifos had survived; it seemed as if, all those years ago, they had changed their minds. Despite the proximity of a huge gold mine at Porgera, they were still living on the mountain. Sometimes, he was told, their calls could be heard, echoing through the mists.

For Benedict, the news came as an immense relief. Now he would have a chance to thank them for their kindness to him as a young man, and to check on their welfare. And maybe, just maybe, he would find out what had happened to Korsai.

In October 2017, Benedict returned to Bisorio with this exact aim in mind. He took no GPS tracker or satellite phone, preferring, as ever, to place his trust in the indigenous people; he knew that, in an emergency, they could get someone out of trouble faster than a rescue team from outside. The same principle applied to navigation. 'Journey with a map,' he once said, 'and all you'll ever come back with is a better version of that same map.' [4]

The team that pushed up into the forested slopes included Michael and another guide, Gibson, the son of Tsogomoi. Despite the gruelling climb and the myriad flies, bees, ants and leeches, Benedict felt exhilarated. He knew that rainforests around the world were being destroyed, but this particular forest, empty but for the Yaifo, had somehow escaped. Although tough to negotiate, it was exquisitely beautiful.

Over the course of a week, the party crossed a succession of watersheds and made frequent detours to avoid landslips and tree-falls; with the help of Tsogomoi's brother, they threw a rope bridge over a gorge and edged their precarious way across. Then, from the vantage point of a ridge, Gibson indicated the unbroken carpet of trees that stretched away into the mist, and announced calmly that the Yaifo must be down there, near the bank of a river. It was time to sing out to them, to herald the arrival of visitors. Eager as they were to meet the Yaifo, no one wanted a repeat of last time.

Gibson let out a soaring, whooping call. It resonated eerily among the trees and faded away. They waited, as rainwater dripped and splashed off the leaves. There was no answering cry. Another guide, Samwell, suggested that the river a mile or so below them was masking the sound to anyone beyond. So they went on, crossing a dilapidated rope-bridge amid another battering downpour, until they saw footprints in the mud.

Just ahead, one or two wooden shacks stood in a small clearing. Nothing moved. The place was apparently devoid of life. This time, Samwell cupped his hands around his mouth and sang out with all the vigour he could muster. They waited again, tense this time, straining their ears and acutely conscious of where they were.

From somewhere close by came a response. *Whoop, wup, whoop!* Human voices and a harsh drumming shattered the stillness. It was intended as a warning, but to Benedict the sound was nothing less than a miracle. These were the voices that he'd feared he would never hear again, and they were approaching rapidly.

These were the voices that he'd feared he would never hear again, and they were approaching rapidly.

Now came the unmistakable clatter of arrows, the smack of strings against bows. A group of Yaifo men, their skin blackened with soot, burst into the clearing and began their noisy, terrifying display, leaping, shouting and rattling their weapons to assert their power, just as they had done 27 years before.

This time, the women were visible. Although they, too, carried weapons in the form of huge knives, there was something noticeably different about the Yaifo's dance. Instead of conveying fear and aggression, it was a joyful celebration. Against the odds, this community of only a hundred or so people had endured. They were confident of themselves and their place in the world.

Benedict's eyes were drawn towards one participant in particular. He seemed older than his companions and needed to pause from time to time to catch his breath. Two red parrot feathers projected from his nose. The man laid aside his arrows and gazed back. He extended a slender hand. It was Korsai. In astonishment and delight, he and Benedict hugged each other repeatedly and called out each other's names again and again.

Afterwards, the whole community assembled around a fire and reminisced, with Gibson acting as interpreter. Other men had already come forward, eager to know if Benedict recalled them as young boys. Benedict passed around photographs from that time, and they laughed as they remembered the dance they'd performed on that first occasion - the dance that had scared Benedict and Tsogomoi so much that they'd felt sorry for them, and decided to stop.

But why had the Yaifo chosen to stay? What had happened to change their minds? Korsai explained: on the morning when they were packing up to leave, they had realised that things weren't so bad up here on their isolated mountain, compared to what was going on down at the mission station. Here, at least, they had their land and their dignity, and now there were no warring neighbours to creep up and attack them in the night. They might even get some sleep!

Like last time, Benedict's stay with the Yaifo passed all too quickly. He was anxious not to burden them longer than necessary; after all, he had been reunited with Korsai and had seen for himself that the Yaifo were thriving, when many other indigenous communities around the world were not. Once again, his departure route took him over the ridge, guided this time by Yaifo men whom he remembered as children.

But after his Yaifo guides bade him farewell and began to retrace their steps, Benedict found his onward route blocked by fighting between two other communities. With Samwell and his remaining companions, he was forced to make a detour in order to find an airstrip where a small plane could be called in. While he was waiting, he became ill with malaria and dengue fever.

Partly delirious, Benedict was nursed by Hewa villagers for a week until help arrived. Somehow, Samwell and his companions had managed to send out an urgent message, and the response came just in time. Once again, Benedict's trust in indigenous people was justified.

In his childhood, Benedict had often wondered what qualities made a successful explorer. Now, well into the 21st century, he writes: 'The era of the "classic explorer" is gone. I was lucky to be a part of that last generation who could just disappear for six months; those other worlds have shrunk down now. But the word "explorer" needs to be redefined, as it's something we all do. The world has never been more accessible – and not just to those with the skills or money. It's an intriguing time when anyone can go off and discover the world around them.'[5]

In the chapter photo, Benedict is shown reunited with Korsai.

[1] Benedict Allen, *Explorer* (2022)

[2] Benedict Allen, interview with Rosie Stancer (December 2023) https://vimeo.com/896186970/03d26d10d2?

[3] Benedict Allen, notes for Jo Woolf (January 2025)

[4] Benedict Allen's website: https://www.benedictallen.com/

[5] *Wanderlust Magazine* (15th November 2018)

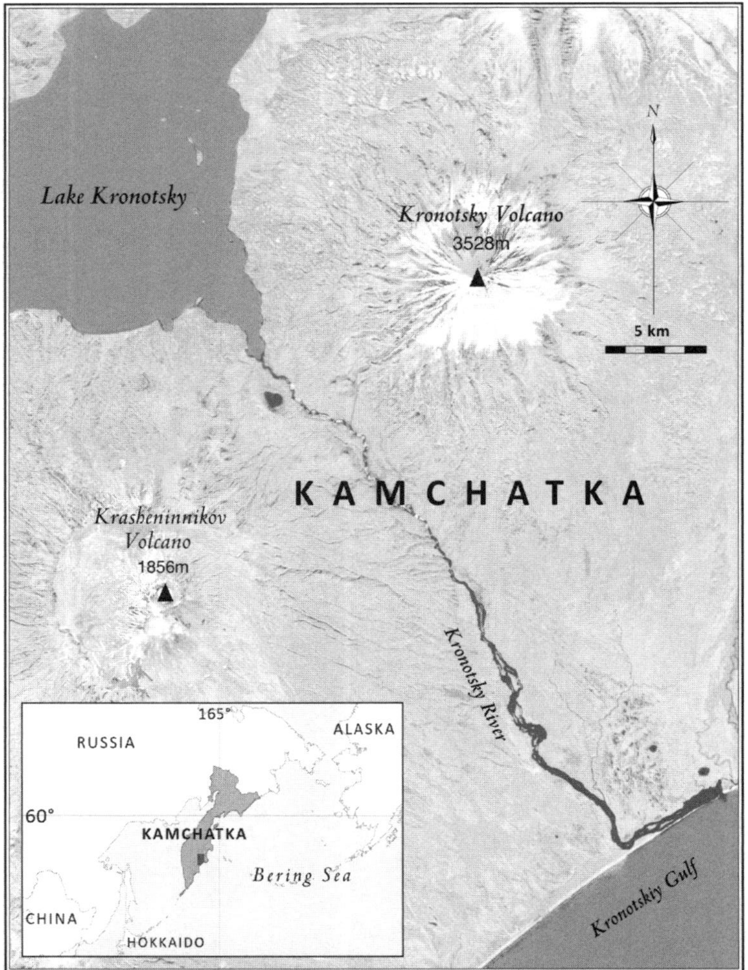
Lake Kronotsky
Kronotsky Volcano
3528m
N
5 km
KAMCHATKA
Krasheninnikov
Volcano
1856m
Kronotsky River
Kronotskiy Gulf
165°
RUSSIA
ALASKA
60°
KAMCHATKA
Bering Sea
CHINA
HOKKAIDO

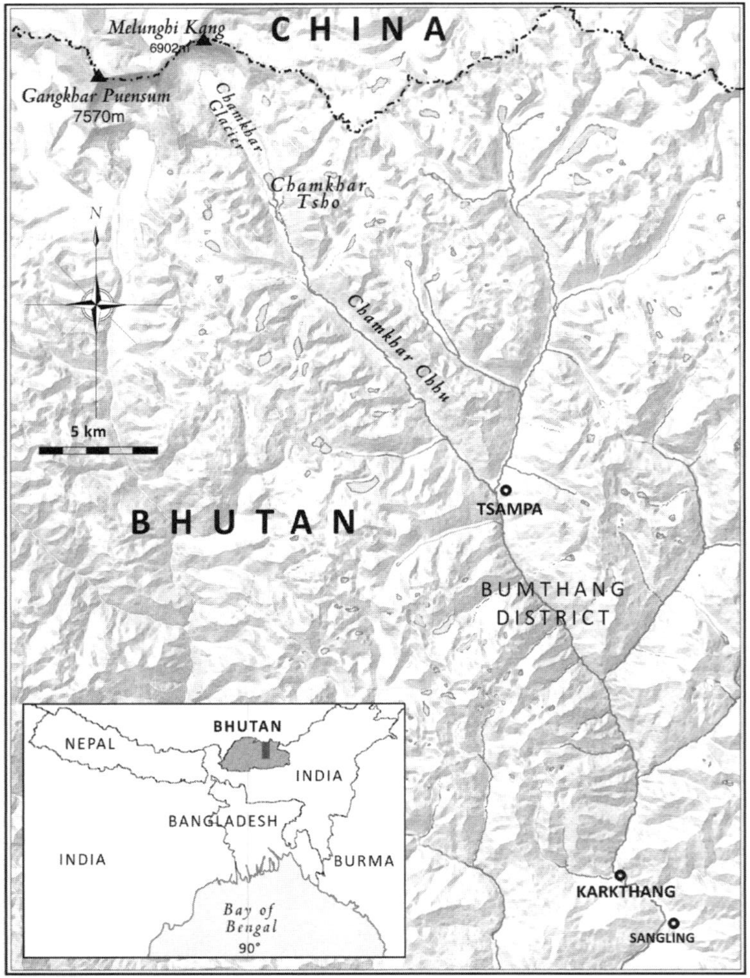
Melunghi Kang
6902m
CHINA
Gangkhar Puensum
7570m
Chamkhar
Glacier
Chamkhar
Tsho
N
Chamkhar Chhu
5 km
TSAMPA
BHUTAN
BUMTHANG
DISTRICT
BHUTAN
NEPAL
INDIA
BANGLADESH
INDIA
BURMA
Bay of
Bengal
90°
KARKTHANG
SANGLING

Sal Montgomery

The Río Claro or Clear River begins its life in the foothills of the Chilean Andes, rushing and tumbling through wooded gorges on a north-westerly course before turning back on itself just south of the city of Molina. Then, swelling and widening with the waters of many tributaries, it joins the Maule River which flows out into the Pacific at Constitución.

In its upper reaches, the Río Claro is the embodiment of a kayaker's dream. Beneath the walls of an ever-deepening canyon, it has sculpted a string of round pools known as *tazas* (teacups), each pouring vigorously into the next in picturesque torrents of white water. From a kayaking perspective, the falls are considered 'clean' - in other words, relatively forgiving for competent paddlers - though some sections require much more skill and experience than others.

Sal Montgomery had only just begun whitewater kayaking when she saw photographs of the Río Claro in a paddlesports magazine. She says: 'It looked like paradise. I remember thinking, one day I want to go there! It really motivated me to keep training every day. I'd be out in all weather, for hours and hours.' [1]

Sal planned to travel to Chile with a group of friends, but for one reason or another each one of them dropped out before departure, leaving Sal to decide if she was prepared to go alone. 'My best friend said, just go, and see what happens. If it's awful, you can come back.' [1]

And so Sal went. Jet-lagged and apprehensive, she turned up at a kayaking hostel where the staff were so welcoming that her nerves evaporated. She got straight out on the water, and by the evening she'd been invited to join a small expedition down the river - a rare opportunity, which she seized with both hands. 'It's such a special place,' explains Sal. 'Once you're into the canyons, there's no way out except to run the river. You've got the adrenalin from running these beautiful waterfalls, but you also know that if anything goes wrong, it's just us. You're fully committed. No matter what else is going on in your life, whatever dramas you might have at home, you can only think and focus on that moment. And when you all come out at the end, there's a great feeling of accomplishing it as a team.' [1]

In Chile, Sal made lifelong friends. She also learned an important lesson. She says: 'It proved to me that I was good enough. That was a life-changing trip for me, because everything changed… I came back as a different person, a different kayaker.' [1]

Sal grew up in Nottingham, where the rest of her family had no real interest in outdoor activities. At school she was withdrawn and disconnected, but once she escaped into the woods she was 'a girl full of curiosity, intrigue, energy, excitement and, most of all, happiness.' [2] Sal's appetite for outdoor adventure tested the resourcefulness of her family, who encouraged her to join local youth groups. 'I was chucked out of the Girl Guides,' laughs Sal. 'They said I was a bit too adventurous!' So she tried the Scouts, 'and suddenly I was around all these people who wanted to get muddy and have adventures. They actually celebrated being adventurous. I really enjoyed that.' [1]

On summer camp with the Scouts, Sal tackled assault courses and went climbing and quad biking, embracing every activity with enthusiasm. Then one summer she tried kayaking. She remembers: 'I'd be the first person on the water, and they'd have to shout at me to come off! I loved it, despite being in a kayak that was too big for me, full of holes and leaking.' [1]

'I'd be the first person on the water, and they'd have to shout at me to come off!'

Although she developed this passion at an early age, Sal assumed that kayaking was something that was only available at Scout camp until she was invited along to the Canoe Club of Nottingham Trent University. 'In Nottingham,' she explains, 'there's a man-made whitewater course which is where the Canoe Club members would train. I had no idea this place existed, but suddenly it became my second home!' [1]

While she was studying physiotherapy, Sal got a job at the whitewater course and took every opportunity to hone her skills. She specialised in freestyle kayaking, which involves surfing man-made waves and hydraulic holes; very soon she was a member of the British Freestyle Kayaking Team. But then she went on a road trip around Europe and travelled further afield to Uganda and Nepal, discovering the wilder thrill of river kayaking.

'When you're paddling a river,' explains Sal, 'you're going on a journey - an exploration - rather than being at one site for multiple weeks in a freestyle competition. There's still a lot of adventure with freestyle, and you can travel to surf massive waves all over the world, but it was the river exploration that really got me excited.' [1] By this time Sal was working as an NHS physiotherapist, and putting in as much overtime as possible to fund her trips abroad. One of these was to Chile, realising her long-held dream of paddling the *tazas* of the Río Claro. She came home knowing that she wanted to make a career out of kayaking.

Soon after her return, Sal had a call from a TV production company. A presenter called Steve Backshall was going to Bhutan to make the first descent of a river, and he wanted Sal on his team. Sal didn't own a television and had never heard of Steve, but she didn't hesitate. 'All I heard was, do you want to go on a first descent expedition? So I immediately said yes!' [1]

But before Sal could even sit in a kayak again, there was an issue that had to be dealt with. Towards the end of her time in Chile, she had badly injured her shoulder. Back home in the UK, she was told it would need reconstructive surgery. For six weeks afterwards her arm would be immobilised in a sling and then she'd have a long period of rehabilitation. Sal was determined to be fit enough to join Steve's expedition, so for two and a half months after the operation, and with the help of her physiotherapist, she devoted every waking moment to re-building the muscles in her arm. Sal's rehab was much shorter than is usually recommended, but her dedication paid off: in September 2018, she flew out to Bhutan with Steve and his team.

Flowing through the Bumthang Valley in central Bhutan, the Chamkhar Chhu enters a gorge just downstream of the town of Bumthang. Then, according to local legend, it falls off the face of the Earth. This goes some way towards explaining why a full descent of the river had yet to be made. Previously, teams of paddlers had by-passed the gorge, fearing that they wouldn't survive the attempt. Now, the kayakers in Steve's team were keen to assess the river and see whether they could get down it safely.

Before they had even arrived in Bhutan, the team had done as much investigation as possible into the challenge that lay ahead. In the absence of reliable maps, satellite images had been examined and the gradient of the river worked out. It was known that the Chamkhar Chhu dropped in altitude from 3,000 metres to 100 metres (9,843 ft to 328 ft) in the space of 100 kilometres (62 miles), meaning that ferocious falls and rapids were almost guaranteed. Rather worryingly, in places the river was so enclosed that no outside rescue could be attempted in an emergency.

After their arrival, Sal and her team-mates spoke to local people about their expedition, and asked whether they were willing to help by sharing information. The story about the river dropping off the face of the Earth cropped up repeatedly. 'We were told that no one went near it,' says Sal. 'So you can imagine how that was making us feel! Luckily, everyone was very welcoming and open to helping us. They were also very concerned for our wellbeing - especially myself, perhaps because I was the girl of the group and the smallest one. I got the impression they wanted to protect me!' [1]

The team learned about a fisherman's trail which would lead them close to a viewpoint. This provided their first real sighting of the Chamkhar Chhu and it confirmed their expectations. 'We could see a lot of white,' says Sal, 'which indicates big whitewater. If it looks white from far away, you know it's going to be full-on. But we could also see beaches and large boulders, which was a good sign that we could get out and inspect.' [1]

In addition to Sal, Steve's team of experienced kayakers included Darren Clarkson-King, James Bebbington, and Chencho, who was from Bhutan. They were backed up by Bhutanese and Nepalese rafting guides, but they knew that on some sections it would not be possible to take rafts down the river. So when Sal, Steve, James, Darren and Chencho took turns to slide their kayaks into the green water and headed downstream, they knew that they would need all of their combined skills and commitment in order to survive.

Sal says: 'I think people consider kayakers to be a bit reckless, just barrelling down a river, but that's not the case at all. When it's a river you don't know, particularly one that's committing or remote or not very accessible and you need to make decisions wisely, you run it in stages. I might be at the front, and I'm always looking downstream at what's next, trying to take the best route for my team to follow me down.' [1]

If it's not possible to see a clear and safe line down the next stage, Sal will signal for everyone to pull into an eddy, usually a pool of calm water behind a boulder. She will then move around the river, trying to see a way through, and if necessary she'll get out and climb up to a vantage point before coming back to give her verdict.

Following this practice, the team on the Chamkhar Chhu repeatedly pulled over before each new stretch of rapids. They were passing through an environment which had never been seen from this perspective by human eyes: slopes forested with pine and spruce, whose fallen needles carpeted the banks in brilliant shades of umber; rock faces that had been sculpted smooth by the water, with the enormous nests of honeybees hanging high above; and water birds - cormorants and shelducks - perching on river boulders, staring at these colourful newcomers in surprise.

At first, the rapids were intense but exhilarating. Then, with rock walls soaring to 200 metres (656 ft) on either side, they approached a scenario that nearly turned the entire expedition into a disaster. Ahead was a fall that was split two ways - on the right, a dangerous siphon which funnelled the main deluge, and on the left, a two-metre-wide chute that flowed right beneath the vertical walls. Sal went first, edging her kayak expertly over the left-hand drop and re-appearing safely downstream a few seconds later. It was down to the others to mimic her course. But when it came to his turn, Steve was momentarily distracted by looking across at the siphon and landed badly at the bottom of the fall. His boat flipped over and he rolled in a desperate, futile attempt to right himself.

In kayaking terms, a hole is a powerfully revolving body of water immediately below a cascade, and once caught in it, even the most experienced paddlers can find themselves trapped. Gasping for air, Steve managed to pull his ripcord and free himself from the boat, then clung to it in the hope that it would be spat out of the infernal cycle. It was not.

Meanwhile, Sal had battled back upstream. Seeing that Steve was fast losing strength, she seized a throw-rope* from her kayak and threw it towards him. She only had one go, because the rope must be coiled back up before it can be re-used, and time was running out. Luckily, Steve caught it. Sal pulled him free and swung him like a pendulum into the main current, and she held on until she was sure that he would travel downstream instead of being sucked back into the hole. Then Darren grabbed him and hauled him out.

Many months later, Steve Backshall's TV series delivered the terrifying power of the Chamkhar Chhu's whitewater straight into thousands of living rooms. Gripped with anxiety, viewers up and down the country found themselves holding their breath. 'They cut it down for television,' says Sal. 'It was actually over five minutes before Steve was out of that hole, which is a very long time to be fighting for your life. I'm thankful every day that he managed to catch that rope!' [1] She wonders - but never for too long - what might have happened if her shoulder hadn't been strong enough, or if she'd never made it onto the expedition at all.

'I'm thankful every day that he managed to catch that rope!'

Ultimately, when the team struck their last camp by the Chamkhar Chhu, they had travelled and filmed a long stretch of this previously undocumented river. There was still more of it left to run - like a treasure, as Steve himself put it, for other adventurers to seek. But elsewhere there were other rivers waiting to be explored, and one of them was arguably one of the remotest on Earth: the Kronotskaya River on the Kamchatka Peninsula.

When Sal was invited to join Steve in Kamchatka, her research, as usual, focused on the climate and the geography: where would they get onto the river, and where could they get out? Were there places to camp along the way? What temperatures could they expect, and did the river level experience a seasonal rise and fall? There were other factors, too. Just getting to their starting point would entail a two-hour helicopter flight over territory with no human settlement; and in summer the region's abundant salmon rivers attract an extremely high concentration of Kamchatka brown bears.†

On a cold, still morning, Sal and the team launched their frost-covered kayaks into a glacial lake beneath the snowy cone of the Kronotsky volcano. 'Everyone was a bit quiet,' recalls Sal. 'We were all in our own heads. I guess we were a bit nervous and anxious.' On expeditions, this is the time when the reality of the location is fully felt and absorbed. 'We are completely reliant on one another. Although we never talk about it, we all know. There's no need to talk about it because we're all very aware of it.' [1]

The Kronotskaya River is serene and calm as it flows out of the lake, but soon it starts to churn and roar with turbulent energy, barging its way through forests of birch and willow. Sal was in the lead as they left the lake behind, and she sighted the first bear within minutes. 'When I came round the corner, I was already sitting up tall and trying to look down the rapid, and instantly I saw a huge bear, standing up on her hind legs, looking straight at me.' [1]

Sal signalled to the other kayakers that there was a bear on the bank, and they all slipped quickly past; then, as she was going over the horizon line of the rapid, she saw another bear, standing in the water this time and scouting for fish. 'That,' she says, 'was the moment when it finally dawned on us that we were really there!' [1]

When she needed to scope out the river by walking ahead, Sal had to take extra precautions. Bears are more likely to attack if they are surprised, and females with cubs are particularly sensitive. Making a constant noise is usually advised, because it gives them time to retreat. But often the only routes that Sal could take along the bank were paths and tunnels that the bears themselves had worn through the bushes. 'I was crawling through these tunnels on my hands and knees, singing, hey bear, hey bear! I had bells on my shoes as well. Then I'd crawl back up and tell the team what the rapid was like.' [1]

Sal didn't come face-to-face with a bear, but she still reckons that they saw about 30 of these magnificent creatures on a daily basis. Each night, the kayakers pitched their tents in vegetation that had been conveniently flattened by bears, and they rigged up an electric fence around the perimeter to dissuade them from coming back. There was other wildlife too, including moose and Steller's sea eagles, which could be appreciated for longer as the the river started to lose its impetus.

On the final day, the rest of the production team took to rafts and joined the kayakers as they ran the last stretch out to the Pacific Ocean. With seals swimming under their boats, they surfed the waves in exhilaration. 'It was a really great feeling,' says Sal. 'There were so many smiling faces. We were at the end of our river, and it was wonderful to share it with everyone. It will always be one of the most special experiences I've had.' [1]

Sal is very conscious of her deep need to be outdoors, immersed in nature. After spending weeks or months with a small group of people in a remote location, surrounded only by natural sounds, she finds the urban world too loud and uses noise-cancelling headphones in crowded places. She adds that life on expedition is simpler and less confusing; she enjoys adapting to extreme environments where she is thrown in - almost literally - at the deep end, and her decisions are high-consequence.

'The outdoors,' she says, 'has given me space to be myself, and to thrive. Confidence, people skills, team work, and a well-developed ability to communicate with a range of people in a variety of circumstances have all come from adventure, but have helped me in every other part of my life also.' [2] And then there is the unequalled experience of navigating wild rivers with a close group of friends, camping under starlit skies many hundreds of miles from human habitation. 'That,' she says simply, 'is such a privilege.' [1]

*A throw-rope is a bag of coiled rope, used as a rescue line in water sports.

† The Kamchatka brown bear (*Ursus arctos beringianus*) is the largest of the Eurasian brown bears, weighing up to 700 kg (1,543 lb).

[1] Interview with Jo Woolf (August 2024)

[2] Sal Montgomery, 'The Importance of Outdoor Education and Adventure', *The Geographer* (Autumn 2021)

PACIFIC
OCEAN
NORTH AMERICA
CANADA
United States
BRITISH ISLES
EUROPE
Russia
ASIA
INDIA
Gibraltar
ATLANTIC
OCEAN
Trinidad
Egypt
Arabia
AFRICA
Equator
SOUTH AMERICA
Congo
St Helena
INDIAN
OCEAN
AUSTRALIA
Falkland Is
St Georgia
ANTARCTIC OCEAN

Anders Holch Povlsen

o

Anders Povlsen first came to the Scottish Highlands as a young boy, on holiday with his family, and he still recalls it as a significant moment in his life.

Having been born and raised in the countryside of Denmark, he already felt connected to nature, and he worked on a farm while still at school. But something about the landscape of the Highlands appealed to him, and this impression remained constant until, as an adult, it developed into a deep interest and a passion for protecting and restoring its natural environments. The catalyst for this passion was the purchase of the Glenfeshie estate, in the south-west of the Cairngorms National Park; as Anders learned more about Glenfeshie and the wider Highland landscape, he realised that here was an interesting opportunity that he didn't want to miss.

With its majestic scenery, Glenfeshie has often been described as the 'Jewel in the Crown' of the Cairngorms. Its beauty had enchanted Queen Victoria, who rode through it on her pony in 1861; it had also inspired the 19th-century artist Sir Edwin Landseer to create his famous painting, 'The Monarch of the Glen', depicting a red deer stag against a backdrop of imposing peaks. Less well recognised, especially in the Victorian era, was the fact that Glenfeshie supported remnants of an age-old woodland that was at risk of disappearing altogether.

After the last glacial period, a natural habitat of temperate forest began to develop in the Scottish Highlands. Around 9,600 years ago, colonising trees seized a foothold in ground newly exposed by retreating ice sheets. A woodland grew up, its open canopy characterised by the tall crowns of Scots pine. Scattered among their characteristic rough, reddish-brown trunks were birch, rowan, aspen and juniper, with alder hugging the margins of rivers and bogs. Beneath their branches, herds of wild cattle and horses grazed, watched by predators such as lynx and wolf.

Exactly when humans first began to cause serious damage to the Caledonian pinewoods is still a matter of debate. The historical ecologist Oliver Rackham writes: 'In early prehistory pine-woods had covered most of the eastern and middle Highlands, but by the time of the earliest documents they were already rare; they had shrunk through prehistoric tillage and pasture or the growth of blanket peat.'[1]

In the 17th century, however, the potential of this forest for shipbuilding was fully recognised. The arboriculturist John Evelyn wrote: 'In Scotland there is a most beautiful sort of Fir growing upon the Mountains… They make our best Mast, Sheathing, &c. heretofore the whole Vessel.'[2] Thousands of Scots pines, prized for their long, straight trunks, began falling to the axe; tar and pitch, also vital for shipbuilding, were obtained from the timber and roots. Trees were extracted on a huge scale via rivers and sea lochs, leaving behind only damaged and disfigured trees, many of which survive today as huge and venerable 'granny pines'.

The fuel-hungry Industrial Revolution found more uses for the timber, and widespread felling continued largely unabated through the First and Second World Wars. By the early 21st century, the natural pinewoods of the Scottish Highlands had suffered hundreds of years of exploitation and misuse through human intervention.

When Anders Povlsen acquired the Glenfeshie estate in 2006, it was still renowned as one of the most scenic areas in the Highlands, with the Feshie River carving a dynamic course beneath the summits of the western Cairngorms. But its woodland cover had been severely depleted, causing wildlife populations to suffer; and to compound the problem, like many other Highland estates, Glenfeshie had long been over-grazed by deer, giving the woodland very little chance to regenerate. The landscape, in many places, showed signs of being seriously degraded. Anders says: 'From the very first moment, it became clear that this was about the landscape… once we really got to work with it, then it was really about trying to protect the integrity and all the biodiversity, and the beauty of it.'[3]

'From the very first moment, it became clear that this was about the landscape.'

The importance of the Caledonian pinewoods has been recognised since the mid-1900s. Rich in biodiversity, they support a number of internationally significant habitats. The rare twinflower opens its delicate pink bells among the heather and bilberry on the woodland floor, and over 360 species of lichen thrive in the clean air. Crested tit, crossbill, siskin and great spotted woodpecker feed on seeds and insects, and the trees provide habitat for pine martens and red squirrels. Capercaillie, which became extinct in the Highlands between 1785 and 1837 through hunting and loss of habitat, were reintroduced in the 19th century and retain a still-precarious foothold.

For Glenfeshie's new owner, it was fair to say that the task of restoring this fragile ecosystem was going to be a challenge. 'Luckily,' explains Anders, 'deep in the glen, there was enough left of the old Caledonian pinewoods that were standing on their last legs, and they were the seed source of something that could burst into life again.' [3] This meant that the regeneration of Glenfeshie's natural woodland was still possible: it just needed other elements in the wider environment to come back into balance.

Inherited from the Victorian era and beyond was the long-accepted practice of keeping red deer populations at high levels to satisfy sporting clients. Very often, deer were fenced out of native woodlands in order to keep them on the bare hill, where they were more accessible for stalkers. In consequence, many animals perished through lack of food and shelter during harsh winters.

At Glenfeshie, Anders realised that a fresh approach was needed, and he was very willing to listen to the advice of conservationists who could see beyond the next shooting season. One of these advisors was Dick Balharry, Area Manager for Scottish Natural Heritage and Chairman of the John Muir Trust. Dick had been familiar with Glenfeshie since 1964; he recalled that, in the 1980s, 'all the vegetation in this Glen, including the heather, was shaved bare by the incessant demand of hungry red deer.' [4]

Dick was an advocate for radical change in management methods. 'If you build a fence,' he said, 'you build a plantation.' [4] What Dick envisioned was a new type of Highland sporting estate, where deer had access to the food and shelter they required, and any hunting would be based in a natural environment that was free of fences. But first, in order to give the natural woodland a chance to regenerate, Dick believed that a drop in deer numbers was essential.

This view was shared by Thomas MacDonell, the Factor for the Glenfeshie estate at the time of Anders' purchase. From his own experience, Thomas knew that erecting exclosures around surviving fragments of native woodland was causing huge suffering to the deer: 'I could see that what we needed to do for a brighter future for everybody was to reduce the deer numbers, grow the woodland, then the native herbivores can live within the woodland and they're not excluded from certain areas by fencing.' [5]

Among other Highland estate owners, managing deer numbers for the sake of the woodland was a practice that did not meet with universal support. Anders recalls that it was a very sensitive discussion, particularly with deer management groups. He says: 'They wanted to keep the deer numbers up, and we knew that actually we have the responsibility to operate differently and to ensure that it's not just about deer, it's about so many other things. It's about the whole system.' [3]

With the deer population reduced to a size that gave the old woodlands a better chance to regenerate, a programme of tree planting was drawn up, encompassing Glenfeshie and its neighbouring estates, also acquired by Anders, of Killiehuntly and Gaick. In the years that followed, hundreds of thousands of Scots pine have been planted, along with other native trees such as birch, rowan, willow, alder, aspen, oak, holly and bird cherry.

This new planting was considered essential because large areas of degraded moorland had an insufficient seed source to allow the natural process of woodland regeneration to happen within the next 10 or 20 years. The plan was designed to help link existing fragmented woodlands and habitats, thereby accelerating habitat restoration. It also takes into account the movements of animals in response to factors such as climate change: 'As the climate warms, a woodland grouse that needs to migrate to higher altitude tree cover cannot if there's simply no higher wooded areas to migrate to, or no wooded avenues to safely travel through.' [6]

Nature progresses at its own pace, and visible changes were not expected immediately. Anders explains: 'It takes time, but if you you don't just look from one season to the next, and instead if you're able to look at it over years, and you have a long-term dream, then it actually happens.' [3]

Nonetheless, some positive signs of natural regeneration were apparent within just a few years. Tree saplings began pushing through the heather, breathing new life into the landscape. A feathery green growth of young pines and birch lit up the glen floor and started to spread slowly outwards, advancing up hillsides that had been bare for generations and ultimately finding their natural tree-line. Recent surveys show that Glenfeshie's original old-growth woodland has already more than doubled in area, extending to more than 1,600 hectares of natural tree growth.

Wildlife is returning in abundance: red squirrels, pine martens, crested tits, jays, crossbills and tree pipits inhabit the woods, while golden eagle, dotterel, ring ouzel and ptarmigan can be seen in the hills. People who live and work in the glen, many of them familiar with a stark post-Victorian landscape, have been astounded by the transformation. And while the flush of new woodland is still in its early stages, already the increase in vegetation will be helping to reduce soil erosion caused by heavy rainfall and floodwater from rivers.

Anders explains that, in Glenfeshie, capercaillie are doing well, and some have even been taken by newly-released wildcats*. 'We don't mind,' he says. 'We're not afraid of predators. I think we might not understand how important predators are in the whole system.' He is also delighted that three newly-arrived families of beavers are in action on the waterways.

Looking back to the difficult early days at Glenfeshie, Anders gives credit to all the members of his team who helped to put a radical new plan into action. He says that Dick Balharry was 'a great inspiration', and he acknowledges Thomas MacDonell for his passion for Glenfeshie's landscape and wildlife. The policy of reducing and managing deer numbers has, he says, been 'a game-changer.'

From the beginning, it was obvious that Anders and his team were thinking far into the future, and considering many generations down the line. Through the establishment of Wildland, a conservation organisation, he aims to restore three estates (including Glenfeshie) in north, central and western Scotland to their former natural splendour. Across these landscapes, a total of over four million trees have been planted and thousands of acres of natural woodland have been regenerated. These, in turn, will provide seed sources for planting in locations where local seed sources have been lost.

Anders is clear about his 200-year vision for the landscape of the Scottish Highlands: 'This is not something that we are doing for ourselves. We will not see the results of this ourselves. This is, on the one hand, much more demanding - but this is also a much bigger dream, about what this landscape has been and what it hopefully should be again.'[3]

Wildland's work is already being recognised. In 2021, it was announced that the 755-hectare Killiehuntly woodland had won the New Native Woodland category in Scotland's Finest Woods Awards. The award acknowledged Wildland's landscape-scale habitat restoration, noting that on the Glenfeshie estate, hard work, collaboration and determination were helping to secure the long-term future of the Caledonian pinewood.

Alongside this passion for restoring the natural landscape is the understanding that humans have dwelt there for millennia, and rural communities have long depended on it for their livelihoods. Anders says: 'We want to see the community do well also; sometimes our job goes beyond trying to do the right thing for the landscape and the habitat. The habitat very often extends into the communities, and you want the community to be happy, and doing well, and supporting these things.'[3] Throughout the estates under its custodianship, Wildland is developing projects which will benefit local economies and visitors alike.

There is also an appreciation of the role of education in promoting a thriving community across the Highlands. Wildland has collaborated with The Prince's Trust and Young Enterprise Scotland, funding both these organisations to set up new programmes in schools and communities across the north of Scotland, encouraging young people to acquire new practical skills and work towards establishing their own businesses.

'Change requires vision and leadership,' said Dick Balharry. 'It is my belief that if we recognise leaders who are willing to change, others will follow.' Speaking in his beloved Glenfeshie in April 2015, he paid tribute to Anders and all the estate staff who had allowed 'a natural living Caledon Forest treasure' to be revived. 'Here,' he said, 'we have an example of how a sporting estate and the public interest can work together.'[4]

The vision embraced by both Anders Povlsen and Dick Balharry recalls the words of 19th-century sociologist and philanthropist, Sir Patrick Geddes: 'How many people think twice about a leaf? Yet the leaf is the chief product and phenomenon of Life: this is a green world, with animals comparatively few and small, and all dependent upon the leaves.'[7]

*Captive-bred Scottish wildcats released by Saving Wildcats, a partnership project dedicated to their conservation and recovery.

1 Oliver Rackham, *The History of the Countryside* (1986)

2 John Evelyn, *Sylva, or A Discourse on Forest-Trees and the Propagation of Timber in His Majesty's Dominions* (1664)

3 Anders Povlsen, interviewed by Jo Woolf (29th January 2025)

4 Dick Balharry, speech at Glenfeshie on receiving RSGS Geddes Environment Medal (18th April 2015)

5 Thomas MacDonell, interview in *The Geographer* (Winter 2023)

6 Wildland: https://wildland.scot

7 Professor Patrick Geddes, farewell lecture in Dundee University, 1919, from Amelia Defries, *The Interpreter: Geddes, the Man and his Gospel* (1928)

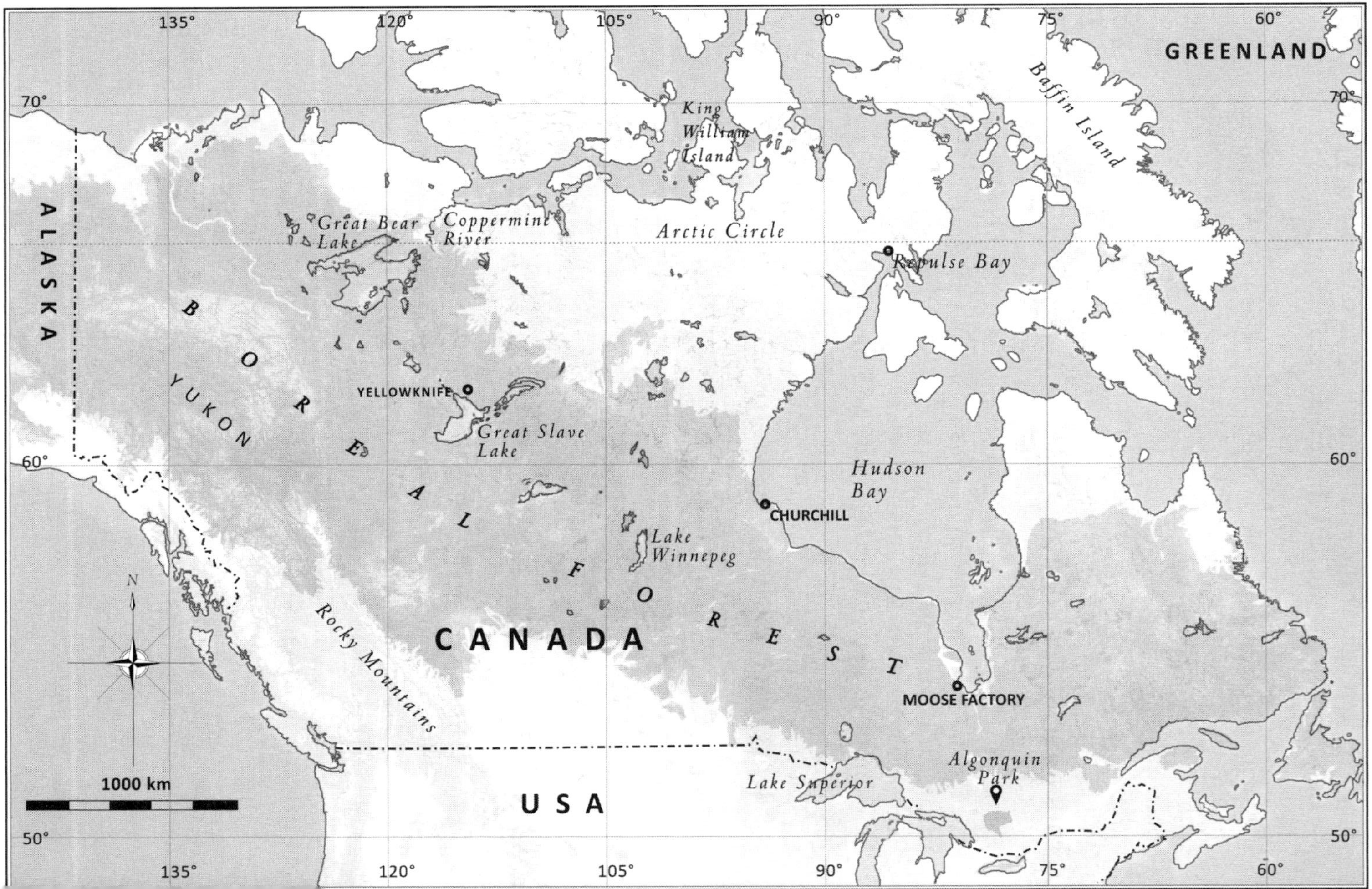
GREENLAND
Baffin Island
King William Island
Arctic Circle
Repulse Bay
Great Bear Lake
Coppermine River
ALASKA
YUKON
BOREAL FOREST
YELLOWKNIFE
Great Slave Lake
Hudson Bay
CHURCHILL
Lake Winnepeg
CANADA
Rocky Mountains
MOOSE FACTORY
Algonquin Park
Lake Superior
USA
N
1000 km
135°
120°
105°
90°
75°
60°
70°
50°

Ray Mears

Ray Mears remembers the first time he camped outdoors. He was still a schoolboy, and with a friend he walked from his home in Kenley onto the local common, on the edge of the North Downs. They didn't have sleeping bags or cooking utensils, and they hadn't taken enough food, but they lit a camp fire and basked in a new sense of independence. From their vantage point, they could see the town of Oxted, where residents were just settling down to watch television. Ray knew where he would rather be.

From then onwards, he was out camping in all seasons. A snowy winter's night became an Arctic adventure. Frustrated by his initial attempts to strike sparks without matches, he deliberately went out in the worst weather and honed his ability to make fire. Sometimes he would follow the trails of foxes, and sometimes he would just sit quietly and watch. Being scared didn't enter his head: he felt that if he had any problems, he could deal with them. 'The woods,' he wrote, 'were swiftly becoming my world.' [1]

Navigation was learned partly from a children's annual but mostly from experience, using a map and compass. He simply set himself a target, and went to look for it. One such target, a natural spring, was extremely hard to find, concealed in a deep hollow that was thickly overgrown with nettles. Ray persevered, and filled his empty water bottle at the spring. He says: 'I think it's really important that you do things like that for yourself - no adult with you, just you and your own self-belief.' [1] But he also acknowledges how lucky he was to grow up in the 1960s and 70s, when children were regularly allowed the freedom to roam and play on their own.

From an early age, Ray learned judo and embraced its principles of courtesy, courage, honour and self-discipline. His teacher, an ex-serviceman who had survived behind enemy lines in Burma (Myanmar) during the Second World War, told him that he didn't need camping equipment, he just needed skills. To acquire these skills, Ray scoured libraries, visited museums, and signed up for courses in the new concept of outdoor survival; he studied everything from expedition preparation to mycology. He didn't just want to survive: he wanted to live in harmony with nature, using bushcraft in the same way that our hunter-gatherer ancestors had used it. By the age of 19 he was practising bushcraft with enough knowledge and confidence to teach it to others. He explains: 'What interests me is becoming more attuned to the environment: understanding it, how it fits together, and what our place in it is.' [1]

Operation Raleigh, the brainchild of Colonel John Blashford-Snell, offered Ray a rare opportunity when he was appointed to the selection team. The project, set up in 1984, allowed young people to join international expeditions with the aim of improving human welfare and protecting vulnerable environments. Ray's own credibility as a bushcraft expert led to the commissioning of a television series, *World of Survival*, in which he visited Indigenous people worldwide, learning about their way of life and seeing them practise skills that have persisted, almost unchanged, for thousands of years - a reminder, he explains, of the humanity of past peoples.

In Australia, Ray witnessed the extraordinary ability of the Anangu people to find underground water in an arid landscape, by following the verbal stories of their ancestors; and in Western Samoa, he learned how the islanders live both from the sea and from the forest, using the coconut palm - which they call the 'Tree of Life' - as a source of food and water (or milk) as well as for roofing, flooring, bedding and basket-making. In Siberia, Ray and his team filmed Evenki reindeer herders, whose lives are almost entirely supported by their animals and the boreal forest in which they live. Whether they were fishing for pike in the lakes or cutting wood from the forest, the nomadic Evenki took only what they needed, a philosophy shared by Indigenous people around the world.

In remote locations, transport for TV film-making is governed by the weather. When heavy snow threatened to prolong his stay in Siberia, Ray was not at all concerned. He explains that, while he is equally comfortable in a wide range of environments, he is especially drawn to the boreal forest. This vast region of trees extends in a wide band right around the Earth's northern hemisphere, from North America through Scandinavia and Russia to Mongolia and northern Japan. 'One-third of Canada,' he explains, 'is carpeted with a mosaic of trees and water, a rich and unique mixture of trees flourishing in the harshest of environments... To understand Canada, you have to understand this forest.' [2]

Boreal forests flourish between latitudes of about 50 and 60 degrees North. In Canada, winter temperatures can drop as low as -53°C (-63.4°F), while summers are comparatively warm - up to 21°C (70°F). For thousands of years, First Nation people have made the forest their home, living in harmony with one of the world's harshest environments. Ray notes that the First Nations have no word for 'wilderness' - perhaps because the forest doesn't seem as alien to them as it does to us; nor is there a word for 'outdoors'. Interestingly, the people who lived outside the forest were said to spend their lives 'on the ground'.

Among the skills mastered by the first inhabitants of Canada's boreal forest was the construction of birch-bark canoes. Made from the skin of a birch tree, spread over a cedar frame and lashed together with roots, these strong but lightweight vessels were ideal for navigating the forest's many waterways, and they were adopted by early European explorers who ventured into the forest to trade for fur. In 2004, when Ray explored Canada's boreal forest for a TV series on bushcraft, he helped Pinock Smith, an Algonquin craftsman, build a birch-bark canoe from start to finish.

Ray had long admired birch-bark canoes, believing that they reflect their builders' intimate knowledge of the natural world, and this was a project that he had always dreamed of completing. However, on putting the boat into the water, it became apparent that there were tiny cracks and holes in the bark, caused by larger than usual lenticels (naturally porous tissues), a phenomenon that was thought to be the tree's response to modern-day pollution. Unable to take the new boat down a stretch of rapids, Ray turned to one which had been made in 1915, and was impressed with its long-term resilience. His experience was a poignant reminder of the importance of cherishing wild places and the species that exist there.

Returning to Canada's boreal forest in 2009, Ray retraced the footsteps of European pioneers who had opened up the country for commercial trade. He chose to live in the manner of the *voyageurs*, who transported furs by canoe at the height of the fur trade in the 18th and 19th centuries. Camping in sites they once used, he replicated their distinctive structure of branches to suspend multiple cooking pots over a fire, and he tried portaging a canoe overland, re-enacting the original opening up of remote creeks and unknown lakes. But he also made it quite clear that these practices had been learned from First Nation people, without whose wisdom very few of these newcomers would have survived.

One of the earliest newcomers to set foot on Canada's shore was Samuel Hearne, whom Ray considers to be among the finest explorers ever to emerge from the British Isles. An employee of the Hudson's Bay Company, in 1769, Hearne embarked on a series of journeys totalling more than 8,000 kilometres (5,000 miles) through largely uncharted territory. His aim was to reach the Coppermine River, which was reputed to be a rich source of copper ore, and he travelled in the company of Matonabbee, a member of the Dene people. By adopting the habits and practices of First Nation people, Hearne proved that it was possible to live off the land on long expeditions, and his detailed reports were an invaluable guide for later explorers.

Ray recalls his own experience of a solo journey by canoe through Canada's boreal forest. After weeks of careful preparation, he watched the floatplane rise above the trees and dwindle into the distance, and realised that his life was now reduced to the simple essentials; yet on the first evening, as he sat by his island campfire, he was aware of a certain unwillingness to relax. Was this, he wondered, how a captive-bred animal felt on being released into the wild? Was it reluctant to leave the familiar safety of the cage?

A gale, whipping up overnight, forced his mind into sharp focus. Believing he could handle the conditions next morning, he packed everything into his canoe and kept it in the lee of the wind as he skirted a lake. Before he could make a three-kilometre (1.8-mile) portage, he had to clear the path of windblown trees. By evening, he felt that he had been truly tested by the elements, but he had renewed his deep connection to the land. As he fished, he watched a moose browsing on sedges by the shore, and beavers carrying branches to their lodge. 'Now,' he wrote, 'I am happy in my own company and I have joined the wildlife around me; we share a secret: the secret of silence.' [1]

Ray still travels to the far north every year, to teach bushcraft and survival techniques. He loves the extreme conditions of an Arctic winter. He explains that, between -6°C and -20°C (21°F and -4°F), the cold is easily managed and the snow squeaks underfoot; but when the temperature drops to -50°C (-58°F), the snow sounds like porcelain and absolute focus is required in order to survive. He writes: 'When, inevitably, we reach a point when no more comfort can be attained, we learn to ignore discomfort and to endure by concentrating on the positive; stoicism is not a trendy idea in the north but a pillar of life itself.' [3]

The benefits of learning to survive in such extreme conditions are extraordinary. Ray describes the 'stark frosted beauty of the boreal forest, the satisfying glow of skin warmed by blood pumped while skiing across a frozen lake, the sound of a Siberian jay chattering in the trees, the ethereal beauty of northern lights as they dance above our campfire where true friendships are forged.' [3] He has spent 40 years teaching people to live and travel in this northern wilderness, and says that no other environment so profoundly enhances the joy of being alive. Some of the skills he teaches have been handed down from First

Nation people, while others were invented by himself; he loves watching the moment when a student becomes aware of their own competence. But, he adds, 'the real teacher is nature herself - I merely facilitate her lessons.' [3]

As a film-maker, Ray wishes he could have witnessed Britain's landscape as it began to warm after the retreat of the ice some 14,700 years ago, with herds of mammoth grazing on grasses and dwarf willows. As trees began to colonise the exposed land, humans would have started to move beneath them. 'How,' he wonders, 'did our ancient ancestors find their way through the un-surveyed forests that flourished after the last glaciation?' [4] He speculates that they must have climbed up into the branches to look for smoke rising above the canopy, a sure way of finding their way back to a campsite.

The ability to make fire is a uniquely human trait, and Ray believes it is the greatest gift we have received from the forest, demanding an equal partnership between humans and trees. Dating the earliest occurrence of fire-lighting is, however, problematic. Charred bone fragments, ash and heat-fractured stone tools found in Northern Cape Province, South Africa, suggest that an early human ancestor, *Homo erectus*, was employing fire at least a million years ago, but it may have originated from a natural forest fire.

As for the act of fire-lighting, some of the oldest equipment yet discovered dates from 50,000 years ago, when Neanderthal communities in France were using flint axes to strike sparks from a mineral, most likely iron pyrite. Interestingly, Ray has encountered desert tribes who had no historical knowledge of fire-lighting, but would carry burning embers with them from place to place. By contrast, the Hadza huntsmen of northern Tanzania carry with them a fire drill so that, wherever they settle to camp, they can take a dead branch from the baobab forest and make a fire that will not only cook their meal but ward off predatory animals during the night.

'Walking in woodland, I genuinely feel that I am surrounded by friends.'

The health benefits of being in woodland, if only for a few hours, have received fresh media attention in recent years. Ray has been aware of them his entire life. 'Walking in woodland,' he writes, 'I genuinely feel that I am surrounded by friends.' [4] The forest was, and still is, our natural home, and it is only in the last few thousand years that we taken our axes and started decimating this precious sanctuary.

In the 21st century, our impact on the natural environment is undeniable. Our ancestors survived because of their knowledge and understanding of nature, and our continued existence depends on the flourishing of our forests. Ray is still hopeful of positive change: he believes that we have more choice over our destiny than ever before. Steps can be taken to protect ancient woodlands, thereby reducing the rate of species loss. Studies can help us understand how pollution is damaging the mycorrhizal network that is so essential to a thriving forest. We can establish greater woodland cover, thereby benefiting biodiversity.

All we need is the collective wisdom and the political will to make the right choices. Ray writes: 'We have tended always to think of what trees can do for us. Why don't we instead ask what we can do for our trees?' [4]

1 Ray Mears, *My Outdoor Life* (2013)

2 Ray Mears, *Northern Wilderness* (2009)

3 Ray Mears, 'Winter, the Season of Wisdom', blog post (3rd November 2023) https://blog.raymears.com/2023/11/03/winter-the-season-of-wisdom/

4 Ray Mears, *British Woodlands* (2023)

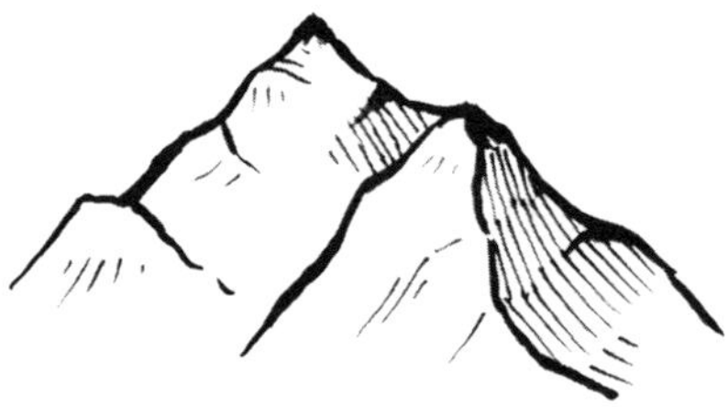

MOUNTAIN

There's no dream that mustn't be dared.

George Mallory, *'Mont Blanc from the Col du Géant by the Eastern Buttress of Mont Maudit'*, The Alpine Journal

Why do you bother climbing? Why do you climb? All I could think of saying is, well, I get grumpy when I don't!

Doug Scott, YouTube interview for Mountain Equipment

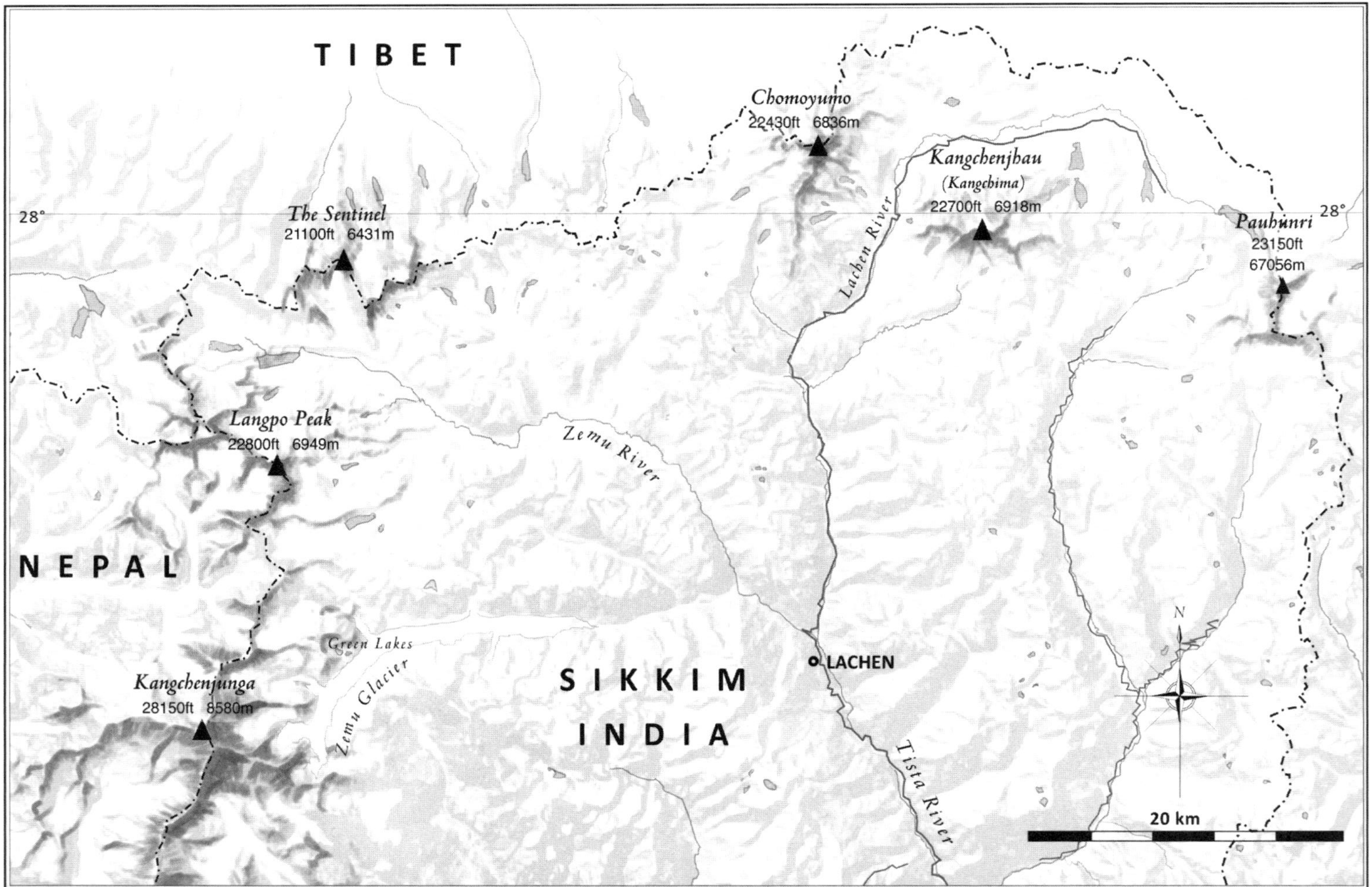
TIBET
Chomoyumo
22430ft 6836m
Kangchenjhau
(Kangchima)
22700ft 6918m
Pauhunri
23150ft
67056m
28°
28°
The Sentinel
21100ft 6431m
Lachen River
Langpo Peak
22800ft 6949m
Zemu River
NEPAL
Green Lakes
Zemu Glacier
LACHEN
Kangchenjunga
28150ft 8580m
SIKKIM
INDIA
Tista River
N
20 km

Alexander Mitchell Kellas

o

On the morning of 18th August 1885, two young brothers, Alec and Henry Kellas, set out from Ballater to walk up Ben Macdui. They carried with them a newly-published guidebook entitled *Ben Muich Dhui & His Neighbours* by Alexander Inkson McConnochie (price one shilling, and bearing a pen-and-ink drawing of the imposing mountain on its cover). In Braemar they bought a loaf of bread, a packet of biscuits and a bottle of lemonade; then they walked up Glen Lui and Glen Derry, and struck west towards the crags and lochans of Coire Etchachan.

As night began to draw in, they scanned the rock fields anxiously for the Shelter Stone, an enormous boulder resting over a natural cavity which had long been known as a 'howff' or walker's refuge. In the darkness of this cave they lit some candles, devoured most of their food, wrapped themselves in their greatcoats and tried to go to sleep. But it was no good: the silence and the cold were too intense. Cutting some heather, they kindled a fire, but it made little difference; at daybreak, bleary-eyed, they went down and washed themselves in the loch. Then they made the ascent to the summit of Ben Macdui, at 4,294 feet (1,309 m) the second-highest mountain in Britain.

The brothers obviously had a passion for the mountains, because instead of giving in to their hunger pangs and beating a hasty retreat back to the warmth and comfort of their uncle's home in Ballater, they carried on along the Lairig Ghru and climbed Cairntoul to within 200 feet (61 m) of its summit, turning back only because it was blocked by snow. Two days later, having slept very little and eaten next to nothing, they were back in Ballater and tucking into a hearty breakfast.

Alec Kellas was 17, and Henry was a few years younger. Long-distance walking was commonplace in the 1800s but this was an epic trek by any standards. Driven by determination and extraordinary stamina, Alec would pursue high-altitude adventure for the rest of his life - and, although he didn't yet know it, he would help others to do the same.

Born and raised in Aberdeen, Alec was the son of Mary Mitchell, whose family hailed from Ballater, and James Fowler Kellas who was Secretary to the city's Mercantile Marine Board. A pupil at Aberdeen Grammar School, Alec went on to study at Heriot Watt College and (possibly) the University of Edinburgh* before progressing to University College London, where he graduated in chemistry in 1892. Then, still at UCL, he took up a post as assistant to Sir William Ramsay, whose ground-breaking experiments in identifying the noble gases helium, argon, krypton, xenon and neon later earned him the Nobel Prize in Chemistry.

While Ramsay's other assistants, Professor J Norman Collie and Morris Travers, received acclaim for their work, Kellas seemed content to stay in the background, performing a supporting role with reliability and thoroughness. Collie and Travers were enthusiastic climbers, but when Kellas went to the mountains he preferred to go alone or with his brother, Henry. He returned many times to the Cairngorms, and in the Alps he climbed the Breithorn, the Finsteraarhorn and Mont Blanc.

In the late 1800s, climbers with a respectable tally of European summits were often looking for a bigger challenge and they found it in the Himalayas. The trouble was that human ambition was racing ahead of scientific understanding. For a start, mapping of the Himalayan ranges was still a work in progress, and there were some regions that were either closed to outsiders or very difficult to access.

Mapping of the Himalayan ranges was still a work in progress.

Then there was the question of increased altitude and its impact on the human body. Compared to Mont Blanc, which rises to 15,781 feet (4,810 m), Everest is a colossus at 29,035 feet (8,850 m). At such heights, air pressure is considerably less than at sea level, and the availability of oxygen to breathe is correspondingly lower, which can cause altitude sickness and hypoxia. Scientists had yet to study this phenomenon, let alone understand it. Meanwhile, climbers expecting to progress with the same ease and speed with which they ascended the Alps found themselves unaccountably beaten back by weakness and exhaustion.

Kellas's colleague in the laboratory, Professor Collie, was making an attempt on Nanga Parbat when his companion, Alfred Mummery, died in an avalanche. Back in London, mourning the loss of his friend, Collie no doubt recounted his experiences to Kellas. Before the accident, he had been shocked at how their performance had been diminished. It raised unforeseen questions. What exactly was causing this response, and what effects did it have on human physiology? Could something be done to prevent it? Although he was unaware of it at the time, Kellas may have been looking at the ingredients of a solution: he was, after all, working with gases. An innovative idea was still a long way in the future, but meanwhile he was slowly nurturing an ambition to see the Himalayas for himself.

A few years later, Kellas left Ramsay's laboratory and took up a post teaching chemistry to medical students at Middlesex College Hospital. It was a rewarding job, and he rented rooms in Regent's Park. He was approaching the age of 40 and could be forgiven for settling into a respectable career. But anyone who had underestimated Alec Kellas up to this point was about to be surprised. Between 1907 and 1912, he took leave from his post for months at a time, disappearing on a series of Himalayan expeditions that he had planned quietly and meticulously. His friend, the climber and photographer John Noel, wrote: 'Kellas did not advertise; few people knew about him. He did not tell the newspapers when he set out to climb a mountain higher than any climber had ever tackled before.' [1]

Kellas's first Himalayan expedition was largely exploratory, but he still cherished hopes of tackling Kangchenjunga, which at 28,169 feet (8,586 m) is the world's third-highest mountain. He got as far as the Zemu glacier, which flows from the east face of Kangchenjunga, but bad weather and dangerous snow conditions forced him to turn back.

In 1909, Kellas climbed Langpo Peak (22,851 feet, 6,965 m); on his third expedition, in 1911, he achieved no fewer than three first ascents: Chomiomo (Chomo Yummo), Sentinel Peak, and Pauhunri. The last of these, at 23,386 feet (7,128 m), was the highest mountain yet ascended by any climber, but ironically Kellas did not know it, because incorrect calculations had placed another mountain, Trisul, slightly higher, and this had already been climbed. The following year, he made the first ascent of Kangchenjhau (Khangchengyao), and survived a 1,000-foot fall down an ice wall which nearly killed him and his two Nepalese companions, Nema and Anderkyow.

Details about Kellas's expeditions were provided in his own modest reports for the Royal Geographical Society and the Alpine Club. A hint of his personality can be gleaned from the accompanying photographs: small groups of sherpas pose, smiling, on ridges and snowy summits, but Alec is always behind the lens. Intensely, perhaps painfully, private, he recorded his achievements by capturing the joy of others.

He recorded his achievements by capturing the joy of others.

Initially, Alec chose Nepalese Sherpas because he found that they were physically well-adapted to climbing at high altitudes. Unlike many of his contemporaries, he treated them well, providing them with tents, warm quilts, and similar clothing to his own. He noticed that they performed well on a largely vegetarian diet, so he tried their own food, prepared from plants gathered in the mountains, and combined the ingredients with his tinned rations.

Just before attempting to climb Kangchenjhau, Alec described his team as being in the best of health and spirits, satisfied with their pay and the amount of work expected of them. Asked if they wished to continue for another month, all agreed without hesitation. They had occasional misunderstandings, usually resolved with a shared laugh. In his reports, unusually for that time, Alec credited the leading Sherpas by name.

In return, Alec found an understanding that probably ran deeper than any he had so far experienced. This gentle, quiet man may have been holding himself apart from close friendships because of a distressing mental health condition that neither he nor any doctors of the early 1900s could satisfactorily explain. He could hear voices in his head, described as auditory hallucinations; they were so clear that he once wondered if a sensitive microphone would allow others to hear them too. In the absence of proper medical understanding or supportive therapy, he must have dreaded the awful diagnosis of 'insanity'.

In the Himalayas, as he camped beneath the vast, star-sprinkled sky with his sherpas, Alec felt free to answer these voices. His Sherpa friends listened without fear or prejudice. In their view, Alec suffered no affliction at all. They told him that they had great confidence in someone who had long conversations with spirits at night.

Kellas's mission involved a lot more than climbing new peaks. On mountain summits, he took samples of air; he measured his own red blood cell count at different altitudes, and he made exact comparisons between his rate of ascent and that of the Sherpas. At the highest levels, he noticed, whether loaded or unloaded, they could easily outstrip him. He wondered whether his poor performance resulted from the reduced formation of oxyhaemoglobin. Although not fully understanding the natural genetic adaptations that allowed Nepalese people to thrive in the mountains, he was pondering possible solutions to the problems of high-altitude climbing.

Kellas, wrote the climber Charles Granville Bruce, was 'the most modest man that ever travelled the Himalaya.' [2] That was entirely possible, but his achievements began to speak for themselves. Having established a reputation as one of the most experienced Himalayan climbers of his day, it was hardly surprising that he should be consulted about the biggest goal of all. Colonel C G Rawling, whose military surveys had established Everest as the world's highest mountain, asked him to draft a proposal for a reconnaissance of Everest in 1915 and a summit attempt, if it looked feasible, a year later.

While the First World War delayed both the reconnaissance expedition and his own plans for a return to the Himalayas, Kellas got to work on a succession of ideas. He devised a scheme for airmen to fly over the Himalayan gorges and survey the terrain so that an approach route to Everest could be developed. Although his proposal was supported by the distinguished mountaineer and geographer Douglas Freshfield, it was received with scepticism by aviators and never put into practice.

There was still doubt about whether humans could survive at all above an altitude of 25,000 feet (7,620 m). In his correspondence with the physician and climber A F R ('Sandy') Wollaston, Kellas suggested that a group of climbers could camp for about a month at a height of 23,000 feet (7,010 m) in order to gauge how their bodies adapted, and he proposed the summit of Kangchenjhau as a possible location. Again, this idea never left the planning stage.

A third project was altogether more productive. With the Scottish physician John Scott Haldane, Kellas conducted experiments in a hypobaric chamber at the Lister Institute in London. Over four successive days, the two men exercised while the air pressure was reduced until it equated to an altitude of 25,000 feet (7,620 m). They proved that short periods of acclimatisation could increase human tolerance to altitude, and that supplementary oxygen could be of significant benefit in improving a climber's performance.

As the leading authority on the effects of altitude on the human body, Kellas was now convinced that an ascent of Everest was physically possible. In a paper submitted to the Alpine Congress in Monaco in 1920, he predicted that Everest could be ascended by someone in excellent physical and mental condition *without* supplementary oxygen if the physical difficulties of the mountain were not too great, and *with* the use of oxygen even if the climb was harder.

Kellas collaborated with the Oxygen Research Committee of the British Admiralty to design oxygen apparatus which he carried on an attempt to climb Kamet (25,447 feet) (7,756 m) with the surveyor Henry Morshead in 1920. The pair failed to reach the summit, but more importantly they proved the efficacy of their new equipment.

An Everest Reconnaissance Expedition was planned for 1921 under the leadership of Charles Howard-Bury, and the Mount Everest Committee invited Kellas to join the team. In May of that year, when they met up in Darjeeling, Kellas arrived late to an official dinner given by the Governor of Bengal. He had walked four miles from a neighbouring village, and his dishevelled appearance made an immediate impact on George Mallory, who was one of the climbers. Writing in amusement to his wife, Mallory likened Kellas to an eccentric alchemist, slight in build and stooping, with round-rimmed spectacles and a long, pointed moustache. But, added Mallory, Kellas was a devoted and disinterested person, and he loved him already.

A long trek lay ahead of them, Accompanied by mules and porters, including four of Kellas's Sherpa friends who had come at his request, the expedition left Darjeeling and crossed the Jelep La pass into Tibet. Before they reached Kampa Dzong they were all suffering from acute stomach complaints, in particular Kellas, who was unable to walk or ride, so his Sherpas carried him in a chair. He maintained a deceptively cheerful demeanour, so on 5th June everyone was devastated when he died.

In February 1922, while speaking to an RSGS audience in Kellas's home city of Aberdeen, George Mallory gave a touching account of how they had laid Alec Kellas to rest within sight of three great mountains that he had already climbed, and the summit of Everest visible in the distance. He explained: 'In him we had lost not only a delightful and helpful companion, but a very experienced mountaineer, of rare enthusiasm…' [3] Sympathies were expressed for Kellas's family; and it would be good to think that one or two of Mallory's listeners went out to the Shelter Stone below Ben Macdui and raised a dram to his memory.

At the time of his death, Alec Kellas and two Nepalese men, Sona and Tuny, still held the summit height record for their ascent of Pauhunri in 1911. Alec's work remained unfinished: later on in the Reconnaissance expedition, he had intended to conduct further experiments on the physiological effects of low air pressure, but no one else had the expertise to take over his role. He would, however, have loved to know of the continued research into breathing apparatus that eventually assisted Sir Edmund Hillary and Tenzing Norgay on their successful summit attempt in 1953. He shared their love of those high places, where he could set his mind free. 'Mountaineering,' he wrote, 'in its widest acceptation is the most philosophical sport in the world.' [4]

* Some of the details of Kellas's education are still unclear.

o

Kellas was proved correct in his prediction that Everest could be climbed without supplemental oxygen if the physical difficulty wasn't too great. In 1978, when Reinhold Messner and Peter Habeler made the first ascent without breathing apparatus, their progress for the last 1,000 feet (305 m) almost exactly matched Kellas's own calculations.

o

1 John Noel, *Through Tibet to Everest* (1927)

2 C G Bruce, *Himalayan Wanderer* (1934)

3 *Aberdeen Free Press* (9th February 1922)

4 A M Kellas, 'A Fourth Visit to the Sikhim Himalaya, with Ascent of the Kangchenjhau', *The Alpine Journal* (May 1913)

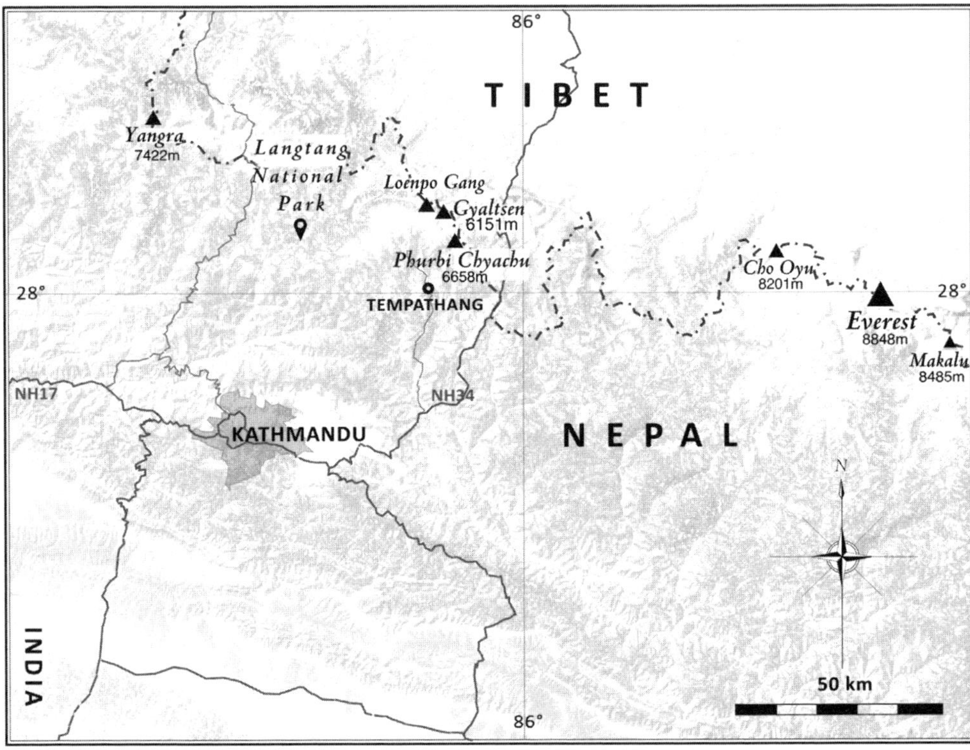

86°
TIBET
Yangra
7422m
Langtang
National
Park
Loenpo Gang
Gyaltsen
6151m
Phurbi Chyachu
6658m
TEMPATHANG
Cho Oyu
8201m
Everest
8848m
Makalu
8485m
28°
28°
NH17
NH34
KATHMANDU
NEPAL
N
INDIA
50 km
86°

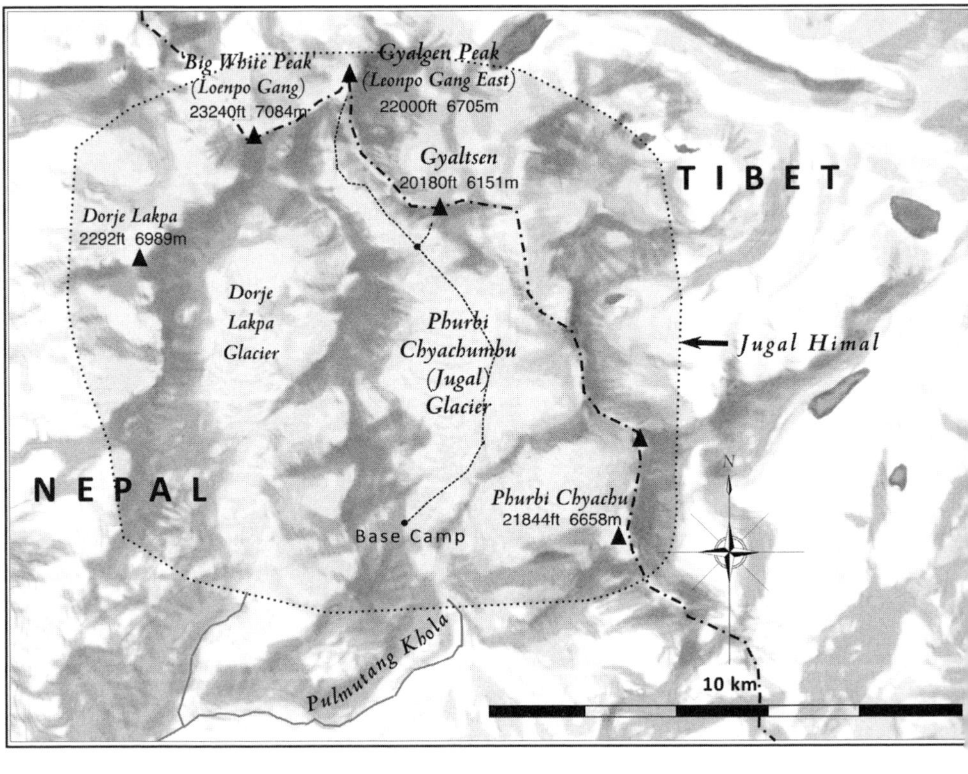

Big White Peak
(Loenpo Gang)
23240ft 7084m
Gyalgen Peak
(Leonpo Gang East)
22000ft 6705m
Gyaltsen
20180ft 6151m
TIBET
Dorje Lakpa
2292ft 6989m
Dorje
Lakpa
Glacier
Phurbi
Chyachumbu
(Jugal)
Glacier
Jugal Himal
NEPAL
N
Phurbi Chyachu
21844ft 6658m
Base Camp
Pulmutang Khola
10 km

Evelyn Camrass

Monica Jackson **Betty Stark**

Their names were Evelyn Camrass, Monica Jackson and Betty Stark. Evelyn was talkative, light-hearted, and fond of giving good-natured advice. Monica tended to be impatient and absent-minded, but she had a strong sense of the ridiculous. As for Betty, she was known for being tough: among the climbing fraternity, it was whispered that she broke into bothies.

In February 1954, Monica gave a talk to the Glasgow branch of the Ladies Scottish Climbing Club. She told her audience that Himalayan climbing was not necessarily the preserve of men. Contrary to popular belief, she added, women had no trouble acclimatising to high altitudes, and could negotiate successfully with teams of Sherpas. These were challenging statements: after all, it was less than a year since the all-male British Mount Everest Expedition had placed Sir Edmund Hillary and Tenzing Norgay on the summit of Everest.

But Monica had done a bit of high-altitude climbing herself, and she knew enough to back up her claim. Afterwards, Evelyn and Betty went to speak to her. They spent the rest of the evening discussing the possibility of an all-women expedition to the Himalayas.

The question of where to go was answered partly by restrictions of accessibility and expense. Tibet was out of bounds, and so were some of the regions controlled by India. Meanwhile, Nepal was granting permits to foreign climbers. The Jugal Himal, a horseshoe-shaped mountain range on the border between Nepal and Tibet, was still largely unexplored. Spurred by curiosity, the three women studied reports written by the climber Bill Tilman, who had tried to get there in 1949 but was driven back by bad weather and the apparent non-existence of access tracks. They wondered why Tilman had called these mountains 'uncompromising', despite their proximity to Kathmandu.

Like a human baby, the Scottish Women's Himalayan Expedition was nine months in gestation. Evelyn, who was a doctor, was put in charge of medical supplies and ensuring that everyone was properly vaccinated. Monica's job was to obtain funding and permits. Betty, whose bothy-breaking reputation turned out to be wholly unfounded, set about ordering climbing equipment and food. It never occurred to any of them to choose a leader: they preferred to rely on mutual cooperation.

One thing they did need to consider, however, was secrecy. Although their ambitions were modest - no Everest-sized mountains were on their hit list - they knew that an all-women expedition was bound to attract a high level of attention, not all of it positive. Betty explained: 'If we slipped up and so got into trouble, there were plenty of people who would say, "These women should never have been allowed out in the mountains on their own."' [1] They therefore agreed to keep their expedition under wraps until the last possible minute. Despite these good intentions, Evelyn, in high spirits, leaked it to most of her friends in the strictest confidence.

'If we slipped up and so got into trouble, there were plenty of people who would say, "These women should never have been allowed out in the mountains on their own."'

Arriving in Kathmandu in April 1955, the women met up with five Sherpas who had agreed to assist them: Mingma Gyalgen* and his cousin Ang Temba; Kusung and his son Bahu; and Chhepela. Mingma had experience both as a climbing Sherpa and a sirdar (head Sherpa) on Everest and Cho Oyu. They hired a group of porters, and collected their 32 crates of supplies. Then, pursued by eager journalists who wanted to know which peak they were going to climb first, they set off on a 50-mile trek to the village of Tempathang, which lies at an altitude of 8,000 feet (2,438 m).

Did these intrepid women wear make-up when they climbed? That was one of the questions debated in some newspapers. Betty made it clear that they did not. Instead, they applied a thick covering of 'glacier cream' (an early form of sun lotion, invented by Swiss mountaineer Franz Greiter) but it was weak by modern standards and failed to stop the intense UV rays from penetrating to their skin. Eventually, even the insides of their mouths got sunburnt, and drinking orange juice was an exquisite torture. Washing was kept to a minimum, since all their water had to be obtained by melting snow. 'We were in a clean world,' wrote Betty, and it was too cold anyhow.' [1]

The path to Tempathang led them through steep, forested gorges. Camping on a grassy shelf high above a valley, they gazed at the snowy peaks of the Jugal Himal on the distant skyline. It was a spectacular but sobering sight. Overhung by formidable cornices, polished slopes of ice rose up into jagged pinnacles. Now it was clear why Tilman had described these mountains as uncompromising. They looked impregnable.

Meanwhile, there was an unexpected hazard. Seasonal burning was sometimes carried out by local people in order to promote fresh grass for grazing, but one fire had got out of hand. In horror, the women watched flames engulf the opposite hillside with the speed of an avalanche, leaving charred trees and ash in their wake. The bank on which they stood had already been burned, but from further up the slope, clods of smouldering turf began to fall around their feet. They moved on swiftly through trees that were still glowing red. A spark caught Monica's shirt, but the flames were quickly put out. Emerging on the other side, they plunged into the icy water of the Nosem Khola river to wash off the dust.

A new team of Sherpas was engaged at Tempathang, and the women scrutinised their maps for the best way into the Jugal Himal. The trouble was that the mountains had all been surveyed from a distance. The map showed three main valleys, each containing a glacier. One of these glaciers might lead up to an unnamed peak which they felt they could climb. But which one? While they were pondering, an old man called Nima Lama came up to them. He knew a path that led up to some some yak pastures, he said. It was rough and steep. He had been up there himself, almost close enough to touch the ice. This was a real breakthrough: they decided to follow it, and see where it went.

Nima Lama was right about the path. After winding through woods carpeted with bright primulas, it ascended steeply and relentlessly for about 3,000 feet. With the increasing altitude, the exertion left Monica, Betty and Evelyn gasping for breath. At about 14,500 feet (4,420 m) they made camp on the floor of a corrie where some stone shelters, once used by yak herders, were grouped around a small lake; they reminded Betty of ruined crofts in the Scottish Highlands. From here, brief reconnaissance expeditions were made: peering through mist that swirled high above the Phurbi Chyachumbu glacier, Evelyn and Mingma glimpsed a 'highway' into the heart of the Jugal Himal. They were jubilant. It was more than they had dared to hope.

The Phurbi Chyachumbu glacier's ice-fall bristled with obstacles, and tackling them demanded all their skill and attention. Slender bridges of ice threatened to give way at any moment, and knife-edged ice walls often guarded an abyss that required a leap of faith. Even when they gained the glacier's upper surface, hidden crevasses lay waiting to trap unwary feet. Monica fell into one up to her waist, and heard shards of ice tinkling like broken glass into the void beneath her feet. Fortunately the climbers were roped up, and she was hauled out safely. Thick mist ruled out any prospect of viewing the terrain from a ridge some 500 feet above their heads, so they turned and retraced their steps with care.

This first spell of serious climbing deepened the sense of trust between the women and their Sherpa companions. Already, camp life had assumed a daily routine. Ang Droma, the 16-year-old daughter of Kusung, augmented their rations by stewing some nettles that she had found growing nearby, and then frying them in butter. Ang Temba, meanwhile, had learned that the women liked to be sure that the water had properly boiled before he made tea, and he would bring the whistling kettle over with a wide grin to prove the point. All the Sherpas were mesmerised by the contents of Evelyn's medicine chest, and she dispensed pills and ointments for conditions that were treatable.

None of the women had expected Himalayan camping to be easy. The necessity of living in cramped quarters for days on end, while coping with headaches and nausea brought on by high altitude, could well have strained their friendships to breaking point. Instead, they mostly laughed at each other's quirks, such as Betty's morning ritual of nose-blowing and Evelyn's habit of hoarding sardines. Monica, perhaps because of her previous experience, was the least affected by altitude sickness: Betty wondered whether her short stature and small frame gave her an advantage, and predicted that the first woman to climb Everest would be a small-boned, petite woman. (She was correct: Junko Tabei, who climbed Everest in 1975, was five feet in height).

Bad weather persisted for several days after their ascent of the Phurbi Chyachumbu glacier, half-burying the tents in snow. When the sky finally cleared, Monica, Evelyn and Betty made it back up the icefall, along with Mingma, Ang Temba, Chhepela, Kusung and Bahu. After their penultimate camp at about 19,000 feet (5,791 m), they struck out for the ridge in bright sunshine. Mingma led, full of excitement, followed closely by Monica; as they neared the top of the ridge, they could see nothing but blue sky and a scattering of rocks. From a natural ledge, the land dropped away vertically for thousands of feet into Chinese-controlled Tibet. Monica remembered the warnings about crossing this border, even by mistake, but here the only way of entering Tibet was by falling in. She and Mingma sat down on the edge, with their legs dangling in thin air, and waited for Evelyn and Betty to arrive.

From this elevated viewpoint, the peaks of the Jugal Himal presented themselves in pristine splendour. Could they tackle one of them? From this high perspective, it was easier to assess the possibilities. One mountain had pure blue ice flanking a relatively rounded summit, but it was also disgorging a difficult-looking glacier. Mingma pointed to it. 'That is a good mountain,' he said to Monica. 'Why don't we climb that?' [1]

'The three rules of Himalayan climbing are reconnoitre, reconnoitre, reconnoitre.' A few days later, this well-known maxim was revolving in Monica's mind as she looked up at the menacing ice-falls and séracs of the glacier they had seen from the ridge. Seen at close range, some of the crevasses had deep walls that were layered in pastel shades of pink, blue and green; another was so wide that a party of four roped climbers was balanced on a slim connecting bridge at the same time. Tempering their eagerness with caution, they made steady progress.

Despite the ever-present threat of avalanche, Mingma stopped to examine some unusual footprints in the snow and pronounced them to be yeti tracks. Evelyn, who was roped up to him, was unconvinced, but a little further up some more tracks were found, running diagonally across a snow corridor. Each print was nine inches long, and - most weirdly - they were at least seven feet apart. Still suspicious, Evelyn asked why there were no toe-prints; Mingma explained that yetis often walked with their toes curled in.

The urgency to reach somewhere safe for an overnight camp made it imperative to move on. Exhausted, and battling a gale that drove needles of powder snow into their eyes, they cast about for a level spot to pitch their tents and beat a hasty retreat into shelter. A thunderstorm broke in the night, eventually rolling away downhill and leaving a brilliant moon hanging in a clear sky. Next morning, the wind was still blowing hard but the weather was fine. It was time to tackle this unmapped and unnamed peak.

For one of the women, however, there was a catch. For several days, Evelyn had been battling the effects of altitude sickness. Nausea, vomiting, and a poor appetite had seriously depleted her energy. Reluctantly, she decided to stay behind.

The final push up the glacier seemed endless. Monica, Betty, Mingma and Ang Temba fought their way up to the shining slopes of pure ice that they had seen from the ridge, cutting steps where necessary. Betty found herself struggling, firstly taking two breaths with every step, then three and four. She told the others that she was holding them back; they assured her she was not. Mingma went further, and with a wicked grin shared some anecdotes about male European climbers that might not have made it into the newspapers. Betty was reassured, although she scarcely had breath to thank him.

Black clouds were gathering in the south: before the last stretch there was time only to swallow some mint cake and strap on their crampons. While Betty and Ang Temba were still a short way behind, cutting steps, Mingma stopped and pointed. The summit! Monica wrote: 'We saw the ice-slope easing off ahead to end suddenly in an unexpected little apex of snow. The sight of it was like a dose of oxygen. I took an enormous breath, said, "All right, come on," and we went up side by side to the summit.' [1]

They had no altimeter, but reckoned they were at about 22,000 feet (6,706 m). On the Tibetan side, the mountain fell away as if sliced by a chisel. A big cornice laced the edge, discouraging them from venturing too close. It was impossible to linger on the summit: the air was so cold that Monica began to lose the feeling in her hands. Amid whirling snow, the party descended as quickly as they could, re-crossing crevasses that looked even more sinister beneath the fast-approaching storm. An ice-bridge gave way under Monica's feet, but as she fell she flung herself forward and struck her ice-axe into the lip of the crevasse. Mingma tried to pull her out with the rope, but her position made it excruciatingly painful and she managed to crawl up on her own.

Back at camp, they stripped off their frozen outer layers and scrambled thankfully into their warm sleeping bags. 'There are times,' wrote Monica, 'when a mug of tea assumes an importance in a climber's life which is not to be underestimated.' [1]

'There are times when a mug of tea assumes an importance in a climber's life which is not to be underestimated.'

Writing about the expedition several months later, Monica, Betty and Evelyn made no claim to have accomplished anything spectacular; they had never hoped to do so, as they were a small party. But they had succeeded in their main aim, which was to explore the Jugal Himal; the ascent of a hitherto unnamed peak was a bonus. They had called it Gyalgen Peak, after their head Sherpa, Mingma Gyalgen; it is now named Loenpo Gang East (22,090 feet, 6,733 m).

In her report for *The Alpine Journal*, Monica paid tribute to the people of Nepal and to their Sherpa companions: 'Of the immense goodwill of all these people I cannot say enough,' she wrote. 'Their courtesy, kindness and enthusiasm for our enterprise endeared them greatly to us. When we left them all at Kathmandu we felt as if we were parting from old friends.' [2]

Monica also shared her opinion that Himalayan climbing was not a relentless slog up to a summit, as it was generally believed to be. Instead, the days that she and her friends had spent tackling the high ridges and glaciers of the Jugal Himal were the happiest and most serene days of their lives. 'Up there,' she wrote, 'each moment was sufficient for itself, and the world pressed upon us not at all.' [1]

*The modern spelling is Mingmar Gyalzen

o

> The expedition's title, the Scottish Women's Himalayan Expedition, does not take into account the inclusion of male Sherpas who played a crucial role in its success; it does, however, reflect the original belief shared by Evelyn, Monica and Betty, that women were capable of organising a serious Himalayan expedition and seeing it through to a safe and successful conclusion.

o

[1] Monica Jackson and Elizabeth Stark, *Tents in the Clouds* (1957)

[2] Monica Jackson, 'The Scottish Women's Himalayan Expedition', *The Alpine Journal* (May 1956)

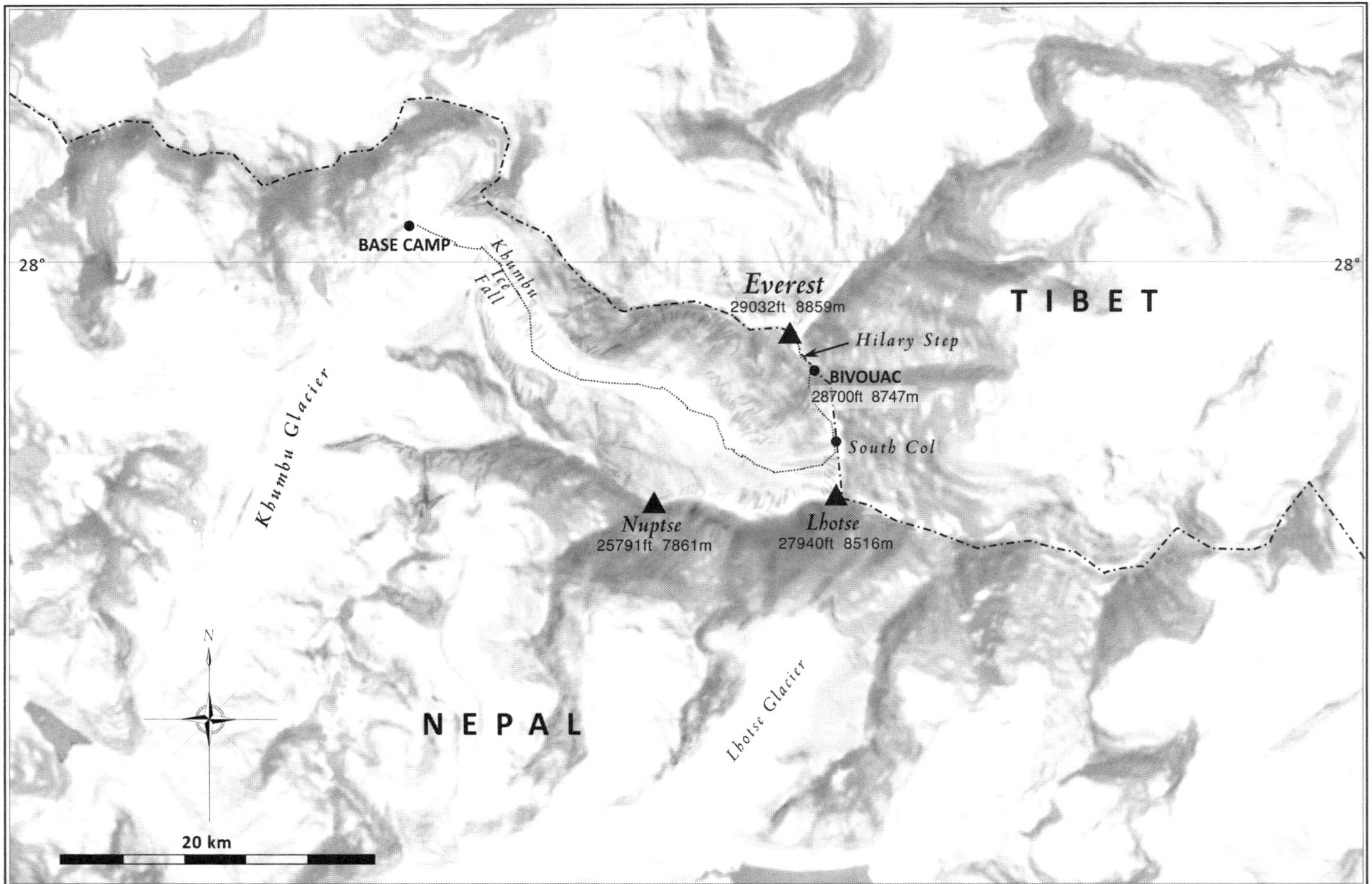
TIBET
NEPAL
Everest
29032ft 8859m
Hilary Step
BIVOUAC
28700ft 8747m
South Col
Lhotse
27940ft 8516m
Nuptse
25791ft 7861m
BASE CAMP
Khumbu
Ice
Fall
Khumbu Glacier
Lhotse Glacier
28°
28°
N
20 km

Dougal Haston **Doug Scott**

o o

There was, thought Dougal Haston, something quite surreal about climbing high on Everest just before dawn. The white ridges gleamed against the deep indigo-blue sky, and the wind - thankfully mild - was picking up twirls of spindrift from the snow underfoot. A mask attached to an oxygen cylinder covered his face, and the noise of his breathing rattled in his ears. Presently he turned to see his climbing partner, Doug Scott, emerge into view, his head silhouetted by the rising sun.

The pair had been awake in their tent since at least 2.30 am, making tea and heating up the remains of last night's corned beef hash before pulling on their boots and climbing suits and strapping on crampons, harnesses and rucksacks. For about a thousand feet they followed a rope they had fixed the day before across a snowfield, but soon they were breaking new ground, alternately wading through knee-deep powder and tackling stretches of rock swept almost bare of snow. They were fit and focused, and they knew that the summit of Everest was within their grasp.

Haston was enjoying himself until he noticed his movements were getting slower. He knew it wasn't his body that was weakening, so he took off his rucksack to look at the oxygen cylinder. The gauge indicated plenty of oxygen left, which could only mean that the system had developed a fault. Scott inspected it too, and was puzzled. An hour ticked by as they fiddled around, trying to persuade it to work. Finally, Scott decided that the only solution was to take the apparatus apart, piece by piece. After all, they couldn't go any further without it. What did they have to lose?

As a rebellious, high-spirited lad growing up in Currie, Midlothian, Duncan 'Dougal' Haston had discovered an extraordinary passion for climbing. He tramped over Scotland's Munros with the local youth club, and at the age of 14 he was guided up Curved Ridge on Buachaille Etive Mor. Until then a capable but uninspired athlete, Haston was exhilarated by the sensation of freedom and controlled risk that came with navigating a soaring wall of rock.

Back at home, climbing became a release and an addiction. With his friends James 'Eley' Moriarty and Jim Stenhouse, he grazed and grasped his way up the rough stonework of local railway and river bridges, with the additional hazard of other kids throwing stones at them from the ground. Hours of study alone in his bedroom convinced his parents that exam successes were imminent, but instead he was reading and re-reading Heinrich Harrer's *The White Spider* and calculating the rate of stone-fall on the North Face of the Eiger.

Who was Dougal Haston? A diligent pupil, with a gift for languages? A tearaway who relished a fight? A fun-loving, irreverent companion? A loner who kept his thoughts to himself? It all depended on whom you asked. He was certainly single-minded: by his late teens, he was tackling the ice-bound winter buttresses of Glen Coe and Ben Nevis under the watchful eye of Jimmy Marshall, a mountain legend who inspired many hopeful young climbers of Haston's generation.

As for his career, Haston had everything planned: he would train as a PE teacher at Glasgow's Jordanhill College, get a job that paid well, and go climbing in the holidays. But before he was due to take the necessary physical tests, he fell off a motorbike at 60 mph and dislocated his shoulder. The chance was gone, so he ended up studying philosophy at Edinburgh University: his reasoning, according to Jimmy Marshall, was that attendance requirements were comparatively light, leaving plenty of free time for the mountains.

Summer jobs in the pea-fields of Kent helped pay for climbing trips to the Dolomites, where some notoriously tough climbs repulsed him, but only temporarily. It dawned on him, while crawling across a dizzying overhang, that he was ridiculously happy. The mountains demanded nothing less than total commitment, and in giving it, he was finding himself. Lying in his bivouac under the starlit sky, he felt a rush of excitement at the unknown paths that lay ahead.

There can be few sights more likely to shatter a young climber's confidence than the North Face of the Eiger. As his train drew into Grindelwald in 1960, Haston described it as the biggest shock of his climbing life. Wind-scoured and ice-glazed, this vertical mile of limestone had already claimed the lives of several climbers, some through its brutal habit of avalanching splintered rock during the afternoon thaw.

Wisely, Haston retreated from his first foray up the North Face, sped on his way by a thunderstorm and volleys of hailstones. A second attempt, in 1962, was aborted when a fellow climber, Andy Wightman, tumbled 150 feet, breaking his ankle and gashing his face. Mentally re-reading *The White Spider* - which is the name of a feature high on the Face - Haston remembered the rail tunnel and its gallery windows that open right onto the rock wall, offering stricken climbers an emergency exit. With Wightman on his shoulders and able to use his arms, Haston got him up there by sheer willpower. So it wasn't until his third serious attempt, in 1966, that he scaled the entire North Face by a new, direct route, joining forces with a German team and naming the route in honour of his climbing partner, John Harlin, who had been killed in a fall.

Reflecting on climbing partnerships, Haston wrote that it was a big responsibility to have another person's life in your hands, and to trust your own life to someone else. Essentially, he liked to be self-contained, but he learned the hard way that life can't be lived in isolation. In 1965, on a dark night in Glen Coe, he was at the wheel of a van that struck three young walkers, one of whom later died. Haston served time in prison and emerged, according to friends, with an even stronger tendency for silent introspection. With renewed vigour, he started pushing himself up ever-harder routes. It wasn't salvation he was looking for, but he knew that in the mountains he could find temporary release.

Haston found work as a climbing instructor and in 1967 he became Director of the International School of Mountaineering, an institution which had been founded by John Harlin in Leysin, Switzerland. Despite the shadows that haunted him, in mountaineering circles he had the dark glamour of a rock star. He partied hard, but months before a climb he abandoned the wild life and focused exclusively on the challenge. His soul was in the mountains. Climbing, he said, had an intellectual side, as well as a practical side. It was a life form, not just a trade. He wrote: 'I've been asked so many times why I climb but in the ultimate analysis I often wonder if I know myself.' [1]

'I've been asked so many times why I climb but in the ultimate analysis I often wonder if I know myself.'

Doug Scott had often pondered a similar question. Bearded, with long hair and round-rimmed glasses, he was the John Lennon to Haston's Mick Jagger. As publicity gathered around his own adventures, Scott wondered whether there was more to his ambition than just a fascination for wild places. Did he desire fame? His decisions to climb, he said, were purely spontaneous. All his life, he had been captivated by tales of explorers. 'For some,' he wrote, 'that impulse to explore is so strong that it transcends reward. They go for the journey's sake.' [2]

Douglas Keith Scott was born in Nottingham in 1941, the eldest son of a policeman who was also an amateur boxing champion. Even before she was married, Scott's mother, Joyce, had been told by a fortune-teller that she would have three sons, the eldest of whom would be in trouble in a shelter, somewhere high up, and that the world would be watching. It was a startlingly accurate prediction, but meanwhile the young Doug was happily getting himself into trouble at lower elevations, running along the roofs of parked cars and digging a tunnel under a railway line with some friends in order to link up two dens on either side of the embankment.

Most of Scott's misdemeanours resulted from an excess of energy. As yet, he had no idea where to direct it. When his school was taken to the Royal Albert Hall to listen to a lecture by Sir John Hunt, leader of the 1953 Everest expedition which saw Sir Edmund Hillary and Tenzing Norgay become the first humans to stand on the summit, Scott fidgeted so much that his teacher threatened to send him out. But a year or so later, on a Scout camp in Derbyshire, he walked to the Black Rocks overlooking the Derwent Valley and was transfixed by the sight of climbers on the crags.

Two weeks later, Scott returned to the Black Rocks with some friends. He was carrying his mother's washing line, while a pal sported his dad's tow rope. The routes they tackled were graded as 'Difficult', but surprisingly none of them came to grief. Initially they learned by trial and error, but shortly afterwards Scott was sent to an outdoor pursuits centre near Buxton, where instructors showed him how to lead a roped climb and arrest a climber's fall by belaying.

Skills could be learned, but wisdom came with experience. Hiking over the Carneddau in North Wales on a relentlessly wet weekend, Scott and his friend, Mick Garside, gave up trying to scale the slippery buttresses of Tryfan and managed to set fire to their tent in the middle of the night. A few years later, on Christmas Eve, Scott and Garside were caught in a snowstorm on Glen Coe's Bidean nam Bian. Wearing just light anoraks, they spent a miserable night in a makeshift snow-shelter and staggered down into Glen Etive on Christmas morning, just as the Glen Coe Mountain Rescue team, led by Hamish MacInnes, were preparing to go and look for them.

After qualifying as a teacher, Scott taught geography, history and games at his old school in Nottingham, but there was a passion for the mountains in his blood and it wouldn't let him rest. Besides, he reflected later, all his life he had fought against inertia, against that temptation to make one more cup of coffee and read the paper. So he climbed in the Dolomites, the Hindu Kush and the Tibesti Mountains in the Sahara. In Yosemite, with the Austrian climber Peter Habeler, he made the first British ascent of El Capitan's colossal Salathé Wall.

On two expeditions led by Sir Chris Bonington - a thwarted

attempt on Everest in 1972, and a successful ascent of Changabang in the Garhwal Himalaya two years later - Scott climbed in the company of Haston. They made a formidable team, coping well with high altitude and possessing exceptional stamina. Mutual respect lay beneath their rough camaraderie: with mock authority, Scott addressed Haston as 'youth', even though he was a year his junior.

In the late summer of 1975, Bonington brought a clutch of talented and experienced climbers together in Nepal. His ambition was not only to put two Britons on the summit of Everest for the first time, but to make a first ascent of the formidable Southwest Face. This was a siege-style expedition - one of the last of its kind before the shift towards lightweight Alpine-style climbing - and the organisational requirements were immense.

A support team of medical staff, communications and equipment officers, drivers, cooks, scores of Nepali sherpas and a BBC film crew had converged on Base Camp in late August. Working methodically, with different pairs of climbers taking turns to lead, successive camps were established up to 27,300 feet (8,321 m). Knowing that they were his two strongest climbers, Bonington consulted Scott and Haston separately: each was keen to climb with the other. He chose them to make the first attempt on the summit.

Mentally, Scott and Haston were utterly committed, but on the morning of 24th September, when Haston's breathing apparatus gave up, they sensed the nearness of defeat. How ironic it would be, after so much human struggle, if mechanical failure forced them to turn back. Scott unscrewed some jubilee clips and inspected the tube that led to the mouthpiece. It was completely blocked with ice. He picked at it with his penknife, and then swung it hard against a rock. Lumps of ice clattered out. Haston put it back on and took a breath: it worked.

Then came a period of agonisingly slow progress, with Haston sinking waist-deep in snow, clearing a space in front of him, moving a few feet uphill and doing the same again. Close behind, Scott stopped him from sliding backwards. They reached the South Summit (a subsidiary peak) at 3.30 pm, heated some water to drink and briefly considered bivouacing there for the night, but the impetus was strong and they pushed on towards the Hillary Step, a 40-foot rock face enveloped in powdered snow. While Haston attacked it with fierce determination, Scott sat down on an arête and took off his mask. With one foot dangling in Nepal and the other in Tibet, he rummaged around in his rucksack and set about changing the film in his camera.

Safely over the Hillary Step, Scott wondered if his oxygen had run out. His mind felt disconnected from his body, and he was observing his actions from somewhere above his own left shoulder. Thinking remotely but clearly, he instructed himself to take a safer line along the ridge. It was good advice: Everest's colossal East Face, rimmed with giant cornices, dropped away for a dizzying 10,000 feet. The ridge was fairly wide, but occasionally one or other of them would break through the crusty snow and stumble. It wouldn't do to stray close to the edge.

At about six o'clock, in calm, clear conditions, Scott and Haston walked side by side onto the summit. Haston was beaming; they hugged each other and turned to look at a sight that left them struggling for words. Across a sea of jagged mountains, the purple shadow of Everest soared for about 200 miles, a vast triangular projection that culminated somewhere in the plains of Tibet. To the east, the setting sun flamed on the slopes of Kangchenjunga, its own lofty peak some 860 feet (262 m) below their vantage point. Silver threads of rivers meandered through valleys already deep in shadow. Majestic, silent, unutterably remote: to Haston, there was a sense of wildness amounting to untouchability.

Scott remembered his camera and fished it out. 'Here you are, youth,' he said to Haston. 'Take a snap for my mother.' Then, after Haston had obliged, 'Better take another one, your glove's in front of the lens.' [3]

Scott remembered his camera and fished it out. 'Here you are, youth,' he said to Haston. 'Take a snap for my mother.'

An hour later, with the sun already set, they turned to go. Abseiling down the Hillary Step was fairly straightforward, but a strong wind was whipping up and a descent to Camp 6 in the dark, with rapidly freezing feet and hands, was no longer an option. However, they had brought no sleeping bags and no food, and their oxygen cylinders were empty. Haston was wearing his down-filled climbing suit but Scott had opted to wear a lighter wind-suit over his clothes. In the dwindling half-light, they used their ice axes to scoop out a snow cave at the South Summit and got out their stove to heat some water from melted snow. It would be a long, excruciating night.

As the temperature plummeted to about -30°C (-30°F), Scott and Haston knew they had to stay awake. They had heard too many stories of climbers falling asleep in these conditions and never waking up. A night at this altitude without bottled oxygen had never been attempted. Hallucinations haunted their reality. Haston engaged in an imaginary conversation with Dave Clarke, who was in charge of equipment;

whenever he started drifting into sleep, some reflex seized him and shook him awake. Scott addressed each of his feet in turn, as if they were people. His left foot felt wooden with frost, so Haston let him warm it up in his armpit. Fighting his ultimate battle against inertia, Scott kept active by excavating the cave until it was big enough for five people.

At dawn, they were still alive. Miraculously, although feet and fingers were slightly numb, there was no frostbite. The only way to get the blood flowing properly was to move. A few hours later, at Camp 6, they were burrowing into their sleeping bags, heating up some food and sharing their news with euphoric team members over the crackling radio. They were the first Britons to reach the summit of Everest, and the first ever climbers to reach it by the Southwest Face. Their bivouac at 28,750 feet (8,763 m) is still the highest on record.

o

> Two more team members, Peter Boardman and Pertemba, a Nepalese mountaineer, reached the summit on 26th September. Exhilaration was tempered by grief at the loss of cameraman Mick Burke, who disappeared en route to the summit, and Mingma Nuru, a porter who died on the Khumbu Glacier on the way to Base Camp.
>
> After Haston's untimely death in an avalanche in 1977 at the age of 36, Scott was one of the bearers at his funeral, his sorrow mingling with helpless anger that their future climbing plans would never be realised.

o

[1] Dougal Haston, *In High Places* (1972)
[2] Doug Scott, *Out and About* (2015)
[3] Doug Scott, from Chris Bonington, *Everest the Hard Way* (1976)

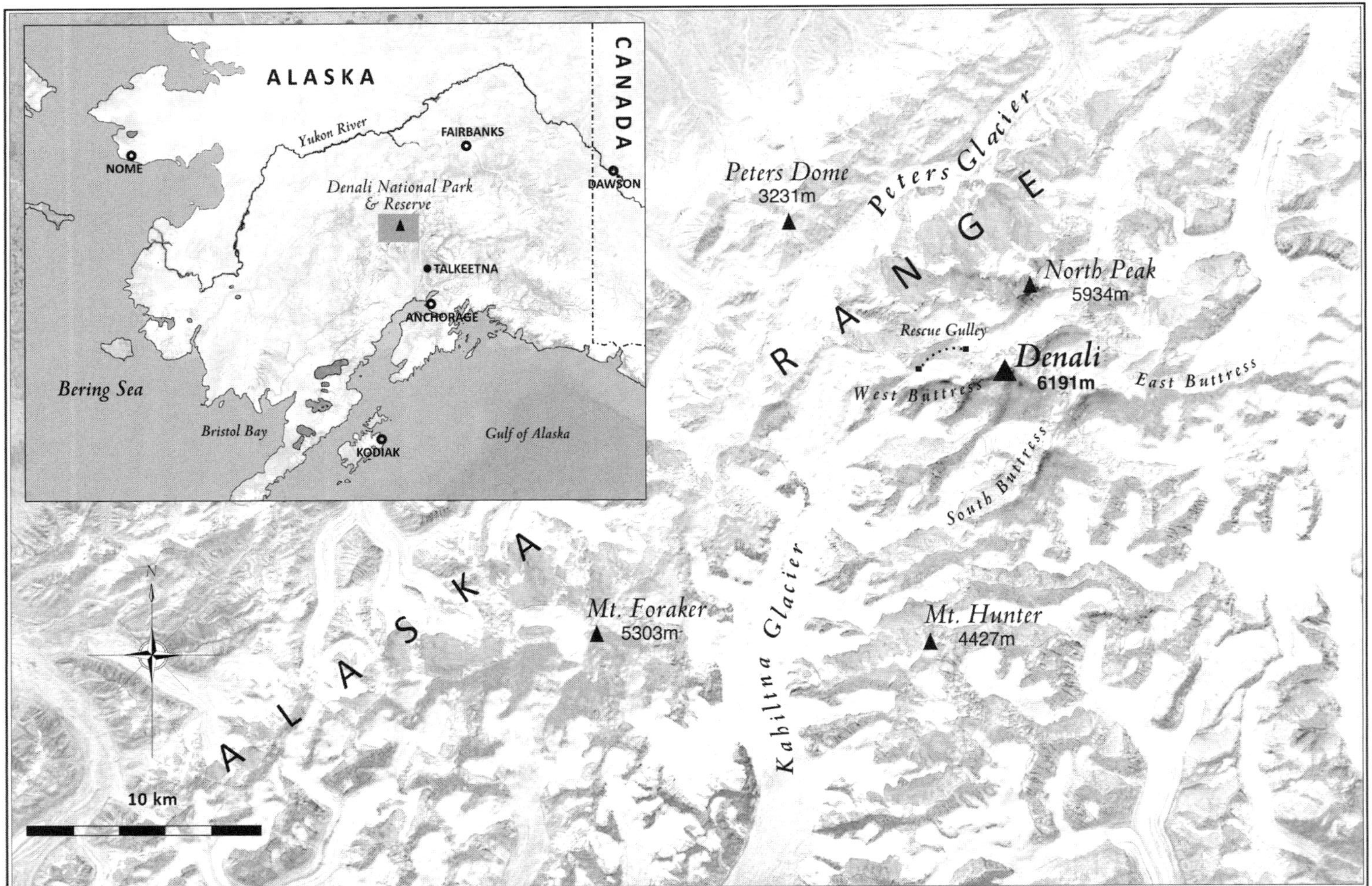
ALASKA
CANADA
Yukon River
FAIRBANKS
NOME
DAWSON
Denali National Park & Reserve
TALKEETNA
ANCHORAGE
Bering Sea
Bristol Bay
KODIAK
Gulf of Alaska
Peters Dome
3231m
Peters Glacier
North Peak
5934m
Rescue Gulley
Denali
6191m
West Buttress
East Buttress
South Buttress
ALASKA RANGE
Mt. Foraker
5303m
Kahiltna Glacier
Mt. Hunter
4427m
N
10 km

Polly Murray

From a small mound of exposed rock at the top of Rescue Gully, an ice-bound couloir near the summit of Denali in Alaska, Polly Murray contemplated the sheet ice that dropped away at a sharp angle from immediately beneath her feet. Far below, the ice was split by a bergschrund or deep crevasse. Polly accidentally dropped her glove and it rolled away, gathering speed as it fell for 3,000 feet down the couloir before disappearing into the chasm.

There wasn't much she could do about it. She strapped on her skis and, together with her climbing partner, Andy Salter, she prepared to follow the route her glove had just taken. She hoped that they'd be able to stop before they got to the crevasse.

Polly and Andy were Telemark skiing, a traditional technique that takes its name from the Telemark region of Norway. Special bindings allow the skier's heels to lift up off the ski, resulting in a smoother and more flowing style. And in 1996, the skis themselves differed noticeably from their modern-day equivalent. Polly explains: 'They were a lot longer, a lot skinnier, and of course you didn't think about getting your edges sharp - they were just a pair of skis!' [1]

No one at that time had tried to Telemark ski from the 20,310-foot (6,190-metre) summit of Denali (which was then known as Mount McKinley). The challenge of climbing up and skiing down was dreamed up in the comfort of Polly's living room. She had been skiing for almost as long as she could walk, but the physical reality of Denali taxed all her skill and stamina, especially with the temperature plummeting to -30°C (-22°F). 'It wasn't the most beautiful ski down,' she says, with a hint of understatement. 'I think we probably survive-skied most of Rescue Gully. We got to the bottom and hugged each other, thinking, oh, thank goodness, we're alive! It was terrifying, to be honest.' [1]

Afterwards, Polly and Andy rejoined the rest of their party and skiied the rest of the way down from their camp at 14,000 feet (4,267 m), pulling their sleds loaded with tents and climbing gear. They had been on the mountain for 21 days. Even the ascent had been arduous, as they had been stormbound for three days at both 12,000 feet (3,658 m) and 17,000 feet (5,183 m). Having been unable to wash for the entire trip, hot showers at the motel in Talkeetna had never been so welcome.

Denali's notoriously changeable weather had also plagued Doug

Scott and Dougal Haston when they made the first ascent of the South Face in 1976. The pair had stood on the summit of Everest the year before, but on Denali they had to draw on all their Himalayan experience to survive. Sudden violent storms raged around their camps, sparking avalanches of powder snow that poured into their sleeping bags. Descending in appalling conditions, they helped save two desperately frostbitten young climbers. Scott later wrote that there is no short cut to climbing at high altitude: the apprenticeship is long and painful.

Polly's own apprenticeship in the mountains began when she was just a child, living among the hills of Perthshire. She was brought up with three brothers, and out of all the siblings Polly recalls that she was probably the most adventurous. Aged five, she climbed Schiehallion with her Mum and older brother, and tackled other Munros with the family. Polly learned to ski in Glenshee, and every winter the whole family would don skis to go off exploring in the woods and hills. 'It was something that came naturally to me,' she says. 'I was able to go anywhere - I loved it.'[1]

'Skiing was something that came naturally to me. I was able to go anywhere - I loved it.'

At school, Polly excelled in Nordic skiing and eventually entered the British Nordic Development Squad, training in Austria before competing in a number of races. But she decided that competitive skiing was not for her, and trained as a ski instructor instead, working in Andorra and then Chamonix every winter, and Telemark ski-touring in her spare time. Aged 17, she walked up to Everest Base Camp with her stepfather, who had led a previous expedition up Everest, and en route she came close to summiting a 21,000-foot (6,401-metre) peak in Nepal. 'That was my first experience of high altitude,' she remembers. 'You get dreadful headaches when you're not acclimatised properly, which I wasn't at the time.'[1]

Each summer, Polly returned to Britain to earn a living from rope access work, which entails abseiling off high buildings in order to carry out repairs. 'I learned to re-point church spires,' she says, 'which was amazing.'[1] By this time, she had climbed Mont Blanc, the Matterhorn and Mount Kenya, and had rock climbed all over the world. One of her employers, who was a climber himself, offered her the chance to go to Denali. At that time, says Polly, people had skied down Denali from about 14,000 feet (4,267 m), but no one had Telemark skied from the summit and down Rescue Gully. After a gruelling climb, she and Andy abseiled a short way down onto a small rocky perch, the climbing rope was pulled away from above, and they were on their own.

Two years later came another opportunity. Polly accepted an invitation to join the British Everest Expedition 2000, which was led by John Barry and included Andy Salter as well as the climber and outdoor equipment retailer Chris Tiso. 'It was a lovely team,' she says. 'Again, I just got lucky. I was already out climbing, I was fit, I was acclimatised because I lived in Chamonix, I had no fear and I was happy to go!' [1]

The team started using small amounts of bottled oxygen from Camp 3, at 24,000 feet (7,315 m). On the last push for the summit, the climbers got up at 3 am and climbed up the Lhotse face to the South Col. After a four-hour rest, they set out again at 10 o'clock that evening and climbed through the night, reaching the summit on the morning of 26th May 2000. 'We were the first team to reach the summit that year from the south side,' remembers Polly. 'We had a massive storm just before reaching the south summit, and another couple of groups turned back, but we kept going. I think that was probably one of the reasons why there was no one else on the summit ridge.' [1]

Thick cloud meant that visibility at the top was down to zero, and there were none of the jaw-dropping panoramic views that greet the luckiest Everest climbers; after lingering for less than 10 minutes on the summit, they started back down. As well as being a member of the first group to climb Everest in the new millennium, Polly had become the first Scotswoman to do so, at the age of 26. It was a huge achievement.

Over two decades after Polly's climb, however, Everest's burgeoning popularity as a tourist destination is having a detrimental impact, not only on the paths up the mountain, but on the natural environment and on the communities that live around its base. Polly says: 'I'm really proud and happy that I did it, but I think it's changed. My experience, personally, was incredible because we didn't have that highway of people trekking to the summit.' She adds: 'Everest is great because it's the highest mountain in the world, but it's not massively technical. I've done other trips that are harder and more rewarding.' [1]

One of those trips came a year after Everest, in the form of an adventure into Arctic waters with Argyll-based sailor and climber Bob Shepton. On board Bob's 33-foot sloop, Polly and her good friend, Tash Wright, joined a small climbing and exploring expedition that sailed to Greenland and Arctic Canada. Polly and Tash were looking for adventure off the beaten track, and their focus for this trip was on Bylot Island, a hundred-mile-long, uncompromising lump of rock and ice at a latitude of 73°16' North.

Located off the northern tip of Baffin Island, Bylot Island is named after the 17th century mariner Robert Bylot, first mate to Henry Hudson on his ship *Discovery*. On an ill-fated Arctic voyage in 1610, Hudson, his son and several crew were set adrift in a small boat by mutineers and were never seen again. In a continuing but fruitless search for the Northwest Passage, Bylot later returned to the Arctic as captain of *Discovery* with William Baffin (whose name was given to Baffin Island) as the pilot.

Bylot Island is bounded by rugged coasts with high cliffs and an interior composed of mountains, glaciers and tundra. In 1963, the mountaineer Bill Tilman made the first crossing of the island; Polly and Tash wanted to do the same, via a slightly different route. They were dropped off on the north coast, with an agreement to meet up with the boat on the south side in seven days' time.

'It was,' says Polly, 'probably the best, most extreme adventure that I've ever had.' [1] While in Greenland, they had been warned that Bylot was full of polar bears, so they agreed to take a rifle in case of emergency. Polly remembers: 'When we first stepped onto the island, the beach was covered in the paw-prints of polar bears - you could almost smell them. For the first few nights in our tent we were really on edge, thinking that they would be sniffing around us.' [1]

But soon the pair had other issues to contend with. Although they had planned their route in advance, a raging river blocked their chosen path onto a glacier, and they were forced to head up another glacier in a different valley. Navigating was an unusually challenging task: their compasses didn't work properly because of the proximity of the North Magnetic Pole*. Polly remembers: 'We basically headed south, used the map and navigated by following the sun and the landmarks.' [1] Often the only way to cross a river was by a natural snow-bridge, which they had to ski across as quickly as possible in case it collapsed. Roped together, Polly and Tash pulled their sleds up through the mountainous interior, thanking their lucky stars that the weather conditions were good.

On the final stretch before they reached the southern coast, Bylot Island threw more obstacles in their path. Polly recalls: 'We started heading down the glacier towards the sea, but after about 165 feet there was a crevasse, and then another and another. Of course, it was where the glacier was steepening. To this day, I don't know how we managed to navigate through them. You'd put your pole down into what looked like snow, and it would disappear.'[1]

Having made it safely to the other side with not even a distant sighting of a polar bear - a fortunate situation which, with hindsight, they could scarcely believe - Polly and Tash settled down to wait for the boat to arrive. During the six-day crossing, their only means of communication with the crew was a VHF radio, which only worked on line of sight. But now they had another problem, because they were running out of food and the boat hadn't yet made the rendezvous. For two days they waited, dividing their last cube of chocolate between them for breakfast on the last morning. The boat, when it arrived, was greeted with joyful relief.

Now living in Perthshire with her husband and their two sons, Polly teaches occasionally at Glenshee ski centre. As a family, they enjoy biking, skiing, kayaking, paddleboarding and embracing the outdoors just as Polly did as a child. Her outlook on life, she says, is to believe in yourself, seize the moment and take the opportunities that come your way.

*Close to the magnetic poles, the horizontal component of the Earth's magnetic field is comparatively weak, meaning that magnetic compasses are less sensitive.

1 Interview with Jo Woolf (July 2024)

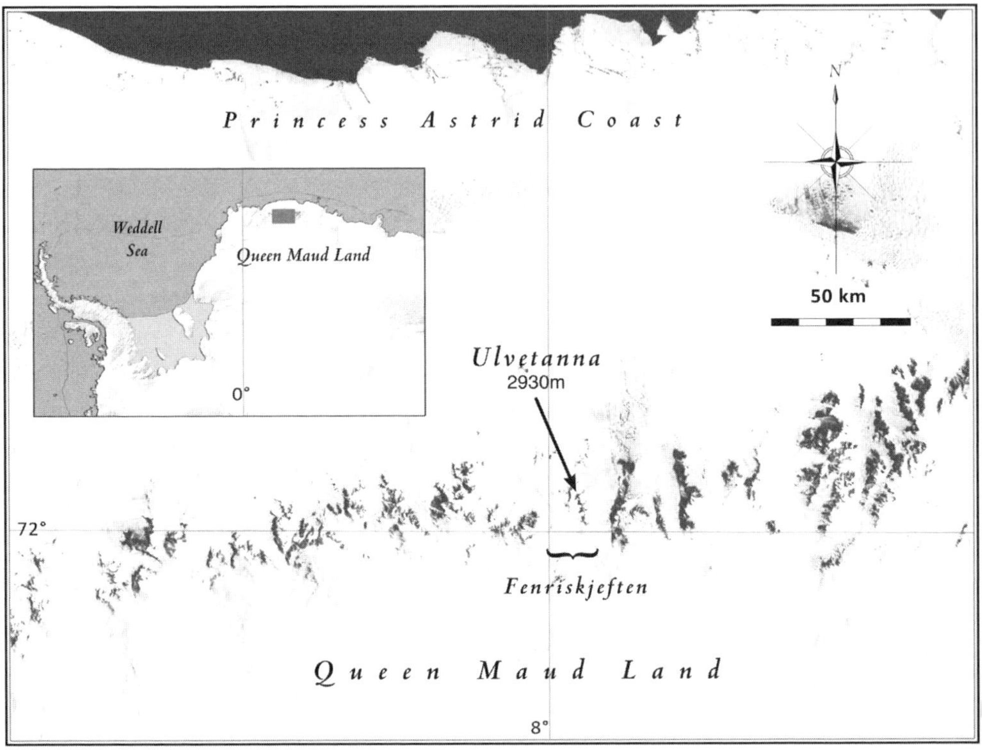

Princess Astrid Coast
N
50 km
Weddell Sea
Queen Maud Land
0°
Ulvetanna
2930m
72°
Fenriskjeften
Queen Maud Land
8°

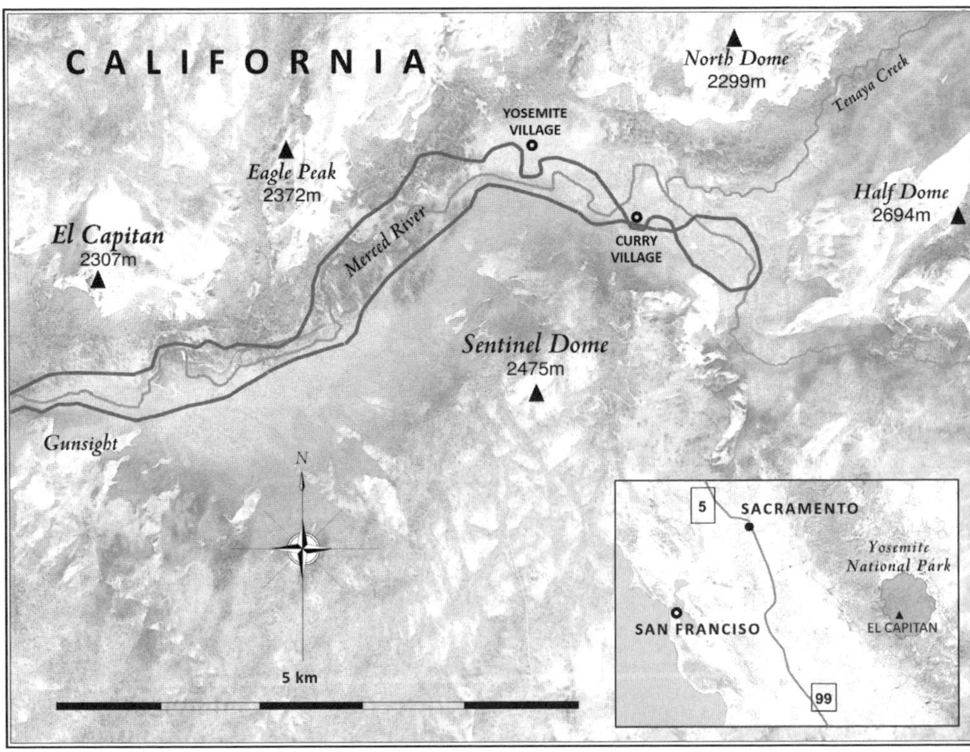

CALIFORNIA
North Dome
2299m
Tenaya Creek
YOSEMITE VILLAGE
Eagle Peak
2372m
Half Dome
2694m
El Capitan
2307m
Merced River
CURRY VILLAGE
Sentinel Dome
2475m
Gunsight
N
5 km
5
SACRAMENTO
Yosemite National Park
SAN FRANCISO
EL CAPITAN
99

Leo Houlding

It was in 1998, while on a climbing and camping road trip through Rogaland, western Norway, that Leo Houlding first saw a photograph of Ulvetanna. A whole book of photographs, in fact, published by Ivar Erik Tollefsen, the Norwegian climber who had led the first ever ascent of it in 1994. Ulvetanna was the most astonishing mountain that the 18-year-old Leo had ever seen, and it inspired a new dream: one day he, too, would stand on the summit.

In Norse mythology, Fenrir or Fenrisúlfr was a ferocious wolf whom the gods feared so much that they chained him to a rock until Ragnarök or Doomsday, when he would break free, devour the sun, and fight against Odin until he devoured him too. The Fenriskjeften (Fenrir's Jaw) massif in Queen Maud Land is named after him. Seen from the air, this extraordinary rock formation with sheer, colossal walls and razor-sharp ridges does indeed resemble the teeth of a monstrous jaw embedded in a desert of Antarctic ice. Rising to 9,612 feet (2,930 m), the highest of these peaks is Ulvetanna, whose name means 'the Wolf's Fang'.

Ulvetanna is so remote that just getting there poses a complex and hugely expensive logistical challenge. For Leo, this meant waiting for nearly 15 years, by which time he had a clutch of other large-scale climbing expeditions under his belt and had forged a strong relationship with his long-term sponsor, Berghaus. From Cape Town, Leo and his five-strong team - Jason Pickles, Sean Leary, Chris Rabone, Alastair Lee and David Reeves - hitched a lift on a cargo plane bound for the Russian scientific base of Novolazarevskaya in Antarctica. Shortly afterwards, a DC3 deposited them and their own small mountain of gear onto the ice about 100 miles inland from the Russian base, and the cheery pilot waved them goodbye.

Leo had done all the preparation himself, from obtaining government permits right down to packing the Christmas crackers and champagne with which they intended to celebrate the festive season. Under a clear blue sky, with the silence of the ice all around them, he and his team stood and stared at Ulvetanna. It looked other-worldly, like a mountain in *Lord of the Rings* or *Game of Thrones*, impressive and intimidating. And now, for six weeks, they would be completely on their own, facing everything that the Antarctic could throw at them. The nearest hospital was 3,000 miles away, in Cape Town. No one would come and rescue them in an emergency. If they ran out of something, there would be no re-supply. Leo remembers: 'Multiple times on the expedition, the thought ran through my head: we are a long way from home - we're so far from home. And you feel it - it's scary.'[1]

'Multiple times on the expedition, the thought ran through my head: we are a long way from home. And you feel it - it's scary.'

Having researched the kind of conditions they might encounter, Leo had chosen their equipment with great care. The durability of the tents was tested within a day of their arrival. News of an approaching typhoon, with winds of nearly 100 mph (161 kmh), was relayed by a stoic Russian voice over the satellite phone. Hurriedly, the team built snow walls as storm defences, then double-poled their tents and roped them together. Luckily, the storm veered away and they were spared the full impact, but three days of white-out conditions and temperatures of -27°C (-16.6°F) made for a chilling welcome, as if Fenrir had just awoken and they were feeling his breath.

In the calm that followed, it was time to move. Half a ton of kit had to be transported three miles by ski and sledge, and then dragged up a 1,000-foot snow slope to reach the base of the mountain. For the first week, in response to the thinner atmosphere at this latitude and altitude, they were plagued with headaches and found themselves gasping for breath in their sleeping bags. Frostbite and sunburn were constant risks. Leo felt a huge weight of responsibility on his shoulders, not just for the safety of his friends, but for his obligations to his sponsors and an extra commitment that he had undertaken, which was to come home with a world-class film of the climb: getting the footage was Alastair's role as a talented adventure film-maker.

But for the moment, the weather was good and the rock quality turned out to be much better than the horror stories Leo had heard about crumbling granite. He had already studied in great detail the route that they intended to take: it was the longest and most spectacular line on Ulvetanna, consisting of a mile-long 'dinosaur spine' followed by the massive vertical headwall of the northeast ridge. He says: 'Sometimes people ask, do you pick these aesthetic lines because they look good in film? No. We pick these aesthetic lines because as a climber, they're the most compelling ones.'[1]

With the challenge of navigating every feature of a new route comes the privilege of naming them. Sean Leary led a 200-foot overhanging pitch* and then, in near-constant daylight, they progressed up a 1,000-foot buttress to the Col of False Hope, battling frozen fingers and descending to camp in a snow-cave when their confused body clocks told them they needed sleep. The 'False Hope' arose from the realisation that their heavy gear and supplies couldn't be hauled up the diagonal route they'd just taken, so they had to pull up all the rope and re-fix it directly down the 1,800-foot north face.

But then came the Stairway to Heaven, a wildly exposed, 200-metre ridge no more than an arm's span in width with vertiginous drops of nearly 2,000 feet on either side. Leo led the climb for four consecutive rope lengths, hand-drilling bolted belays in between. Technically, he didn't find it too difficult, but the exposure was phenomenal and there was no protection: he felt as if he was walking the plank into outer space. It was lucky that there was so little wind. Re-climbing it for the sake of the film a few days later, he was quietly telling himself, 'Don't fall now, or you'll die!'

On a flat terrace about the size of a golfing green which they called The Plateau of Great Expectations, the climbers gathered to survey the sky-terrain above their heads. There was another 1,000 feet of vertical rock, but that wasn't the end of it: the headwall simply obscured what lay beyond. They descended to base camp to welcome in the New Year, and then, suitably rested and rejuvenated, they went back up to the Plateau, using the north face ropes to haul up sufficient supplies for two weeks. They pulled up the fixed ropes behind them, removing their escape route and committing to the headwall. On the plateau they attached two portaledges (suspended tents) which would now be their shelter until they either reached the summit or were forced to turn back.

The headwall wasn't giving up without a fight. Leo used handwarmers in his chalk bag to fend off frostbite in his bare fingers as he battled with steep corners and splitter cracks†. Up to now, the team had been free-climbing, but that was about to change. Another storm hit them full-on, chasing them back down to the portaledges where they sat for three days until the wind subsided. Then a very different scene met their eyes: the Stairway to Heaven was covered in fresh snow, and a safe descent that way looked impossible. In these much colder conditions a mix of free climbing and aid climbing was now their only option, swathed in down jackets, Gore-Tex, big gloves and alpine boots.

Moving at the rate of two pitches per day, it was a battle-hardened team that passed slowly through the Slot of Despair - a flared 'chimney' through a huge horizontal roof - and up towards the summit of Ulvetanna. Blizzards returned, and the temperature dropped to -35°C (-31°F). Beards and eyelashes were coated with rime; hands and feet had to be massaged to keep the blood flowing. Now, it was a raw battle against the mountain and the elements. Fenrir, the wolf of Ulvetanna, was baying and venting his wrath. But when they hauled themselves onto the tiny summit pinnacle, it was the turn of the climbers to howl in exhausted triumph. For a fleeting moment, the clouds broke apart and bathed them in golden light.

'It's those hard experiences,' says Leo, looking back, 'which are the most formative and memorable ones. When I catch up with old friends, we never discuss those days when everything went according to plan. It's always the major epics, where you snatched victory from the jaws of defeat.' [2]

Growing up in the Lake District, Leo had an adventurous nature that propelled him into escapades at a young age. For many years, his favourite 'toy' was a discarded rope-rigging kit that had once belonged to a telegraph-pole worker. It consisted of ropes of varying lengths with eyelets at the end, some industrial-size carabiners, and a sit-harness. Using the support of some sycamore trees behind his house, he set up all kinds of rope swings and zip-wires that served as rudimentary abseils and Tyrolean traverses.

On days out walking the fells with his family, Leo and his dad would scramble up Pinnacle Ridge above Ullswater, where a short section demanded proper climbing skills. Rock faces took on a special attraction: Leo realised that the excitement of adventure could be found right there, just a few metres off the ground. With a climbing mentor, Malcolm 'Pike' Cundy, he practised on the 30-foot crag of Hoff Rock near Appleby, completing ever-trickier routes until he could climb the hardest one solo.

Aged seven, Leo went on a two-week hiking expedition in Turkey's Taurus Mountains with his father and uncle, during which he learned to rationalise his fear of exposure in high places and focus on the secure placement of his feet. He was fascinated by the problem-solving element of climbing, and enjoyed using his strength and suppleness to overcome vertical obstacles. He started sport climbing on artificial walls, although traditional climbing on natural rock always had a greater appeal. He attended talks by climbing icons such as Sir Chris Bonington, and marvelled at the Aladdin's cave of equipment in the home of Doug Scott, who lived nearby. Aged 11, he became the youngest person (at that time) to climb the Old Man of Hoy.

In Pike's mountaineering library, Leo found a book on Yosemite by the photographer Ansel Adams. The scale of the mountains there boggled his mind. El Capitan, Pike told him, was 3,000 feet (914.4 m) high from base to summit, and took most teams a week to climb. Leo compared this with the highest mountain in the Lake District - Scafell Pike, at 3,209 feet (978 m), and that was measured from sea level. He realised that he needed to go to Yosemite and see the magnitude of 'El Cap' for himself. In his late teens, after a couple of summers tackling the slate slabs and sea cliffs of North Wales, he felt ready for a big adventure.

By the 1990s, the Yosemite Valley had long been attracting seasonal gatherings of risk-loving itinerant climbers. Many were rising to international fame, and they were revelling in the journey. They welcomed Leo as one of their own: he'd barely time to recover from his initial awe at the geological colossus of El Cap before he was dangling ecstatically from a rope on the notorious Porch Swing, with 3,000 feet (914.4 m) of fresh Californian air beneath his feet. Then, with energy and enthusiasm, he set about exploring this new vertical playground, searching for the 'Braille trail' of connecting hairline cracks and matchstick edges that form a free climb.

Aged 18, Leo became the first Briton to free-climb El Cap. In 2008, with Sean Leary, he free-climbed both El Cap and the Half-Dome in a single day; and in 2010 he put up a new free-climbing route on El Cap, which he called The Prophet. But arguably his most memorable experience on that iconic granite buttress was the night when he and Jason Pickles were caught in a sudden violent storm and had to climb back up in order to survive.

They'd reached the top at sunset and began the East Ledges descent route when a deluge of tropical intensity turned the rock face into a powerful waterfall. Clad only for a summer day, and with static electricity humming like a transformer in their metal climbing gear, they struggled back up through raging watercourses and heaved themselves into the relative safety of some scrubby bushes just below the summit. After a night of near-hypothermia they chose the long way round and hiked for eight miles through fresh snow, arriving back at their campsite relatively unscathed. Their appearance was greeted with relief by a rescue team that was about to go and look for them.

That night, remembers Leo, was one of the most powerful and formative of his life. Things could have gone badly wrong, but through focusing decisively on survival he'd found clarity of mind and untapped reserves of strength. With hindsight, he knew that the emergency could have been avoided, either through climbing faster, starting earlier, or taking more protective clothing. 'Nobody,' he cautions, 'should go

Scottish Women's Himalayan Expedition: camp on the Phurbi Chyachumbu glacier *Monica Jackson, courtesy of Simon Jackson*

Polly Murray on Everest
John Barry

Dougal Haston climbing the Hillary Step on Everest, 24th September 1975
Doug Scott Photo Library

Ulvetanna in the Fenriskjeften range, Antarctica
Alastair Lee

Portaledges occupied by Leo Houlding's team on Ulvetanna
Alastair Lee

Leo Houlding's team climbing Ulvetanna
Alastair Lee

Townspeople in Ha'il, Saudi Arabia, taken by Gertrude Bell in 1914
Special Collections & Archives, Newcastle University

Dandan Uiliq, showing excavated Buddhist shrine
Illustration from 'Ancient Khotan' by Aurel Stein (1907)

Rory Stewart and Babur in Afghanistan

Amur Al Wahaibi, Mohamed Al Zadjali and Mark Evans
Crossing the Empty Quarter/John Smith

Mohamed Al Zadjali, Mark Evans and Amur Al Wahaibi, in the Empty Quarter
Crossing the Empty Quarter/John Smith

Brahim Ahalfi, Alice Morrison and Addi Bin Youssef on the Draa Expedition
Alice Morrison

Examining a haul on board HMS *Challenger*, from *The Sea: Its Stirring Story of Adventure, Peril and Heroism* by F Whymper, 1887

out into a serious storm with the intention of having an "epic" - but when it does happen, you'll never forget it!' [2]

Leo's record of spectacular climbs extends from the Alps to South Africa and from China to Venezuela. He has helped recreate George Mallory and Sandy Irvine's bid on Everest in 1924, and has fought his way through the near-impenetrable forests of Guyana in order to free-climb the formidable 'Prow' of Mount Roraima. He has snow-kited for 1,000 miles (1,609 km) across the Antarctic to scale the 6,628-foot (2,020-metre) Spectre in the Gothic Mountains, and in 2023 he made a second expedition to Asgard on Baffin Island, forging a new route which he called 'Loki's Mischief'. As a climber, he says that the aesthetics of a mountain are an important part of its appeal: 'I enjoy climbing really beautiful, striking features and lines on steep, spectacular mountains, especially on the big, complex expeditions where you're spending a lot of time and money and energy to get somewhere.' [2]

From a young age, Leo's philosophy of reaching for his wildest dreams has taught him to deal with fear and risk in all its guises, to accept the possibility of failure, and to develop the self-belief that can make great things happen against considerable odds. 'If there isn't uncertainty,' he says, 'then it isn't adventure.' [2] He considers himself among the luckiest people, because early in life he discovered something that he was both good at and passionate about. For Leo, climbing is not an option but a necessity. It isn't always about the hardest or the highest: sipping coffee while watching a glorious sunrise from a desk-sized ledge is just as memorable as the last gruelling pitch to a hard-won summit. And the people he shares his adventures with are just as significant as the places themselves.

'If there isn't uncertainty, then it isn't adventure.'

Inevitably, over the years, Leo has suffered the loss of friends. Like all professional climbers, they knew and accepted the risks they were taking, but the grief still struck him hard. When he and his wife, Jessica, started a young family, Leo doubted for a while whether he could reconcile his passion for adventure and its inherent risks with the responsibility of parenthood. But he wants to inspire his children to reach their highest potential and to know that they can achieve anything that they set their hearts on, and he believes that the best way a parent can do this is by setting an example. He remembers the upbringing that he had himself, a fine balance of care and freedom, encouragement and trust, along with a readiness to experience other countries and cultures.

As a family, Leo, Jessica, Freya and Jackson have climbed in Slovenia, in Wyoming, in the Alps and in Norway; they stand together on high summits, but they also enjoy hiking, camping, fishing, kayaking, skiing and all the fun of being together on an outdoor adventure. Children, Leo believes, are more resilient than they are often given credit for. 'If you try and keep it fun, and we do have various tricks to do that, it's quite remarkable what they're capable of.' [2] In 2020, Freya, aged seven, became the youngest person to climb the Swiss peak of Piz Badile unaided, while a photo of three-year-old Jackson on the same summit, safely roped and beaming happily, went viral on social media. He had been carried for most of the way on his mother's back.

'I'm not a very materialistic person,' Leo explains. 'Memories are the valuable things for me... getting out into the wildest, least accessible corners of the Earth and interacting with the landscape, playing on a massive scale. And one of the nice things about climbing is that you don't just go and look - you're not in the mountains for a couple of days, you're there for weeks, even months. You see different facets and you experience them at different times of day, in different seasons. They really are incredibly powerful places.' [3]

* pitch: a section of a climbing route between two belay points, limited by the length of the climbing rope.

† splitter cracks: long cracks in a rock face in which fingers, hands or feet can be wedged.

1 Leo Houlding, interview for Trek & Mountain (2014) https://www.youtube.com/watch?v=aMuWsq2Gz6o

2 Interview with Jo Woolf (July 2024)

3 Interview for RSGS (2014)

DESERT

So you leave the sheltered close, and, like the man in the fairy story, you feel the bands break that were riveted about your heart as you enter the path that stretches across the rounded shoulder of the earth.

Gertrude Bell, *The Desert and the Sown*

After one such journey the desert has caught you. If you will, you, the so-called teacher of men, shall be taught yourself things which are never learnt in the hurry, the bustle, the crowd, and the jazz of twentieth century life.

Mildred Cable and Francesca French, *Something Happened*

Mediterranean Sea
SYRIA
TRIPOLI
LEBANON
BEIRUT
DAMASCUS
Syrian Desert
ISRAEL
AS-SUWAYDA
NAZARETH
Jabal al-Druze
SALKHAD
AMMAN
JERUSALEM
JORDAN
PETRA
AL-QA'IM
Tigris
TIKRIT
Euphrates
BAGHDAD
Ctesiphon
IRAQ
Al-Ukhaidir fortress
Nafud Desert
SAUDI ARABIA
HA'IL
Red Sea
MEDINA
N
600 km
36°
38°
40°
42°
34°
32°
30°
28°
26°
24°

Gertrude Bell

Gunfire. The sound pierced the stillness of the night. With it came the distant strains of men's voices, raised in a battle song. Gertrude Bell, still awake in her camp, went quickly outside to listen.

Her travelling companions were seated around the embers of a fire: three muleteers from Beirut, a cook from Jerusalem, and a kinsman of the sheikh of the Da'ja tribe, who was acting as her guide. Their mules had been tethered close by. To the north rose a low volcanic hill that was crowned by the ruins of an ancient fortress, and here and there drifts of snow glittered beneath a bright full moon.

The commotion was coming from the fortress. As Gertrude stood and gazed, she saw the flames of a beacon leaping against the dark sky.

Immediately, she knew what it was about. This was a call to arms. Strong feeling lay between the scattered tribes who roamed the highlands of the Jebel Druze (Jabal al-Druze). The day before, on arriving in the town of Salkhad, Gertrude had heard about a recent raid carried out by Sakhr and Howeitat tribesmen. Five thousand cattle belonging to the Beni Hassan tribe had been stolen, and the Beni Hassan were eager for vengeance.

Gertrude had been disappointed to miss the raid, but now she could witness something just as exciting. She approached a local Druze man, a guest of her party. 'Is there permission to my going up?' she asked. He replied: 'There is no refusal. Honour us.' [1]

Together, they climbed the hill. Around the castle walls had gathered a crowd of young men and boys, armed with swords and knives. They were chanting phrases of a song that shocked Gertrude with its savagery. She wrote: 'Each line of it was repeated twenty times or more until it seemed to the listener that it had been bitten, as an acid bites the brass, onto the intimate recesses of the mind.' [1]

The participants formed a circle and joined hands. A young warrior strode up to each in turn and shouted a challenge: 'Are you a good man? Are you a true man?' Each answered with a cry of readiness. Then he came to Gertrude. He raised his sword above his head. 'Lady!' he said, 'the English and the Druze are one.' She replied: 'Thank God! We, too, are a fighting race.' [1]

Martial spirit had infected her, had overridden her logical mind.

When they turned and ran headlong down the hill, she ran with them. Only when they reached the valley did she stop and let them rush past, reflecting that this kind of activity might not sit well with her respectable image as a desert traveller. She went back to her tent and, as she put it, 'became a European again, bent on peaceful pursuits and unacquainted with the naked primitive passions of mankind.'[1]

When they turned and ran headlong down the hill, she ran with them.

No one, seeing Gertrude growing up amid the privileges and comforts of nineteenth-century wealth, could have predicted that she would choose to wander alone in one of the most dangerous regions of the world. Her grandfather, Sir Isaac Lowthian Bell, was a politician and industrialist whose iron and steel works on the River Tees supplied bridges, rail track, submarine cable and steel rope to Britain and its expanding Empire, thereby establishing the Bells as the sixth richest family in England. Gertrude's father, Hugh, took over the Bell industries and built a family home at Redcar, with a private railway platform in the garden so that he could step on a train and go to work.

Gertrude, her brother and three half-siblings developed a love of the outdoors and (in Gertrude's case) a loathing of discipline. She saw off a succession of governesses and was eventually sent to Queen's College, London, a residential school for girls. She made such brilliant progress that in 1886 she was admitted, aged 17, to Lady Margaret Hall, Oxford; only two years later, she emerged with a First Class Honours Degree in Modern History, the first woman ever to do so.

Although she was presented to Queen Victoria as a debutante and plunged into the obligatory whirl of balls and parties, Gertrude failed to secure the dazzling match that was expected of her. Well informed and outspoken, she had strong political views and in intellectual arguments she took no prisoners. Secretly, she would far rather have been galloping across the moors at home, riding astride instead of side-saddle when she could get away with it.

Gertrude's first experience of the Middle East came when she was invited to stay in Tehran with her uncle, Sir Frank Lascelles, who was a British diplomat. Each morning she would take a horse and ride out into the hills, gazing rapturously at the landscape and marvelling at her new-found sense of freedom. Everything enchanted her: the Persian language (which she had quickly mastered), the poetry, the rose gardens in the city, and the ruined towers and temples that spoke of lost civilisations.

She was ready to fall in love, and when she met Henry Cadogan, an embassy secretary who could match her in wit and sharp political debate, she thought her happiness was complete. But their marriage plans were quashed by Gertrude's father, who told her that Henry's income was insufficient (privately, he believed Henry to be a gambler). Broken-hearted but surprisingly submissive, Gertrude returned home; perhaps to distract herself through physical exertion, she threw herself into some severely challenging expeditions in the Alps.

But in 1905 she was back in the Middle East, at the invitation of a friend in Jerusalem. This time, she was more mature, more confident, and she knew what she wanted to do, which was to ride out and explore the country on her own terms, with no one to control her. Hiring mules and muleteers, she was gone for weeks and then months at a time, becoming bolder and more adventurous, speaking fluent Arabic with her guides, learning to ride and direct a camel, protecting her skin from sunburn with swathes of fabric, and steeling herself to drink muddy water from waterholes.

After an invitation to join a Bedouin sheikh in his tent, Gertrude learned the hard way about the traditional courtesies of the desert. For an hour or two she drank coffee and listened to musicians but then, unaware of giving offence, got up and said goodbye. Back at her own camp, her guides told her she had committed a serious solecism: she should have stayed for a meal because a sheep had been killed in her honour, and she should also have taken a gift. So she returned with apologies, stayed for dinner, and presented the sheikh with a pistol, wrapped in a pocket handkerchief.

As one expedition followed another, either crossing the Syrian and Arabian deserts, tracing the Euphrates River to Baghdad or examining the ancient palaces of Ukhaidir and Ctesiphon, Gertrude's natural air of command and her growing knowledge of political and tribal intrigues impressed themselves on the minds of desert dwellers who hosted her. Learning that her style of travel was just as important as her destination, she indulged her taste for luxuries, packing a Wedgwood dinner service, silver candlesticks and hairbrushes, lavender soap, Egyptian cigarettes in a silver case, a comfortable chair and a folding canvas bath.

Pistols and binoculars were carried as potential gifts; her wardrobe included practical riding skirts, sweaters, fur coats, hats, veils and parasols along with couture evening dresses and fine lawn blouses. Hidden beneath her silk petticoats, a pistol was strapped to her leg. She trailed an aura of greatness: after she had left, one of the Druze chiefs enquired of his villagers, 'Have you seen a queen travelling, a consuless?'

Over time, Gertrude's desert journals developed into an repository of priceless information. Moving from one town or one camp to the next, she noted what was being said in certain circles, and used it to interpret what she heard in others. Through her network of family contacts in embassies and other government offices, her observations were passed on, via letter or word of mouth, to the diplomats and statesmen of the day, and her unrivalled understanding helped to inform decisions at the highest political level.

As for the real reasons behind her extraordinary journeys, Gertrude's principal driving force was her own curiosity. She was fascinated by people, their customs and beliefs and allegiances, and communicating with them gave her great satisfaction. She planned her expeditions according to what she wanted to see; the sense of freedom was intoxicating. But her most daring journey, which she began in December 1913, was undertaken for a different reason.

Aged 38, Gertrude was helping to investigate archaeological remains around Konya in Anatolia (part of modern-day Turkey), when she met a British diplomat and former military officer named Charles ('Dick') Doughty-Wylie. At tea with Dick and his wife, Judith, Gertrude found that she and this highly decorated war veteran had much in common, namely a fascination for Eastern cultures and religions and a passion for poetry. In the years that followed, friendship grew into mutual desire. But Doughty-Wylie, aware that they were courting scandal, drew back from the public admission that he feared would destroy his wife, while still declaring his love for Gertrude from a distance.

Fearful of losing him, Gertrude dreamed up a new enterprise that she hoped would alleviate her pain and regain his interest. She planned a journey to Hayyil (Ha'il), a city at the geographical centre of northern Arabia and stronghold of the Rashid dynasty, whose long-standing conflict with the Saud dynasty had had a powerful impact on the political landscape of Arabia. Getting to Hayyil involved an arduous trek of 1,600 miles (2,575 km) through potentially hostile territories in the Syrian and Nafud deserts; a handful of European travellers had attempted it, and some had paid with their lives. Adding to the danger was the imminence of war: Britain was lending support to the Saud dynasty, while the Ottoman Empire, which was strengthening its ties with Germany, supported the Rashid.

British diplomats warned Gertrude not to attempt the journey, but she ignored them. She welcomed the risks. Part of her wanted

Doughty-Wylie to worry about her. She told him that she would write a detailed journal of her travels and send it to him in instalments. In Damascus, she hired a 'chief of staff', an experienced desert traveller called Muhammad al-Marawi, who helped her acquire guides, camels and camel-drivers. Right from the outset, she had plenty of drama to relate.

Only a short distance from Damascus, a horseman came galloping towards them, brandishing a rifle. Other riders appeared, rough-handling Muhammad and forcing Gertrude off her camel. They ransacked her bags and proceeded to strip the men of their weapons. Only when a young boy in Gertrude's party burst into tears and claimed that he recognised the riders, and had in fact been their guest, were they obliged by their strict code of conduct to return the bounty and offer a guide of their own tribe over the relevant territory, as tradition demanded.

It was winter, and heavy overnight rain turned the ground into a quagmire. Lighting a fire to cook food and dry clothes risked attracting more unwelcome attention. Gertrude gave up hope of being able to wash. Shivering in her tent, she gazed at the stars and pictured her family as they celebrated Christmas. For four days she was detained by suspicious Turkish soldiers, who refused to allow her to go any further until she made a written declaration absolving them of responsibility for her fate. From the nearest telegraph station they brought a message from the British ambassador to the Ottoman Empire: carry on at your own peril. She had, as she put it, cut the thread. Now she was truly on her own.

But Gertrude had never lacked courage. Whenever they came across a Bedouin encampment, she rode straight up to the sheikh's tent to introduce herself. She understood the importance of correct conduct and the need to impress tribal chiefs with her own status; her demeanour, and her retinue of camels and baggage, must have spoken volumes. Some chiefs were more welcoming than others: one or two saw opportunity, and bargained ruthlessly for items such as her telescope and a gun. But if they underestimated her, they had to pause for thought when, seated on cushions in their tent, she discussed with them the meanings and origins of ancient texts and recited poems in Arabic that she herself had translated.

As her caravan moved across the barren terrain, Gertrude navigated by compass because there were so few landmarks. Water was scarce; the going was slow and treacherous, with dunes of deep, soft sand rising to steep precipices on the other side, rimmed with cornices that were liable to collapse. The camels were exhausted, and despite her best intentions Gertrude became depressed. She was still writing her journals for Doughty-Wylie, but at the same time she wondered if this grim pilgrimage was actually worth it.

With relief, some two months after she had set out, Gertrude sighted the battlements of Hayyil in the distance. Lit by the rising sun, fringed with plum and almond trees and set against a backdrop of blue mountains, the city presented a vision worthy of an Arabian fable. Observing tradition, she sent outriders ahead and presently some representatives from the city rode out to meet her. She entered Hayyil like a true queen of the desert, flanked by a party of sword-bearers.

She entered Hayyil like a true queen of the desert.

While her men unloaded the camels, Gertrude was led through shadowy, whitewashed chambers into apartments that, she was told, would be her lodging. Two women awaited her there: Lulua, with thickly plaited hair and strings of rough pearls, was the caretaker, while Turkiyyeh had belonged to the harem of the previous Amir or ruler of Hayyil. But the present Amir, a 16-year-old youth, was away with his men on a desert raid; no one knew exactly when he would return. Only Ibrahim, his deputy, had remained in Hayyil. He received Gertrude warmly, but she heard nervous whispers. Her arrival was giving cause for concern.

Then came the news that she was forbidden to leave. In effect, she was under house arrest. The phenomenon of a woman travelling alone had aroused suspicion, and she was only allowed out into the courtyard. Perhaps they thought she was a spy for the British government: she would, of course, have reported what she saw and heard on her return, but she had no official role. Having paid off her men, she had run out of money, and the customary letter of credit she had brought with her, obtained in Damascus, was being ignored. Ibrahim told her that Fatima, the Amir's powerful grandmother, refused to recognise it. In desperation, Gertrude sold most of her camels, reflecting that a return journey with no money and a much-reduced cavalcade was asking for trouble - assuming, of course, that she was ever allowed to leave.

Meanwhile, somewhere out in the desert, the young Amir was ordering the murder of his companions. Among them was Zamil ibn Subhan, the Amir's uncle, who acted as his Regent and advisor; Zamil had been advocating peace with the Rashids' rival, Ibn Saud, a policy with which the Amir strongly disagreed. Zamil's brothers were murdered too. In Hayyil, no one was yet aware of the massacre, but the chances of Gertrude's release on the Amir's return looked decidedly slim.

In growing impatience, Gertrude talked to the women of the family. She was allowed to meet the harem, carefully cloistered and arrayed in rich brocades and jewels, and realised that they were the inheritance of successive victors who had seized control through violence and bloodshed. Mudi, the Amir's mother, had been married to three consecutive rulers; she was fascinated by this strange visitor in a lace evening gown, who lived freely as a privileged independent traveller. For both women, this meeting was a revelation. Each glimpsed, as if through a small window, a world that neither of them could fully comprehend.

After eleven days of captivity, one of Gertrude's attendants appeared at her door, holding a bag of money. Her credit had been honoured. There was no explanation. He told her that she was free to go.

Typically, Gertrude didn't just ride off into the sunset. Before she left, she explored the palace, climbed up to the rooftops, and went out onto the surrounding plain to take photographs. Then she said an affectionate farewell to the women who had shown her kindness. She never discovered who had granted her freedom: perhaps it was Mudi, bestowing a gift that she herself would never possess.

For emerging safely from an extraordinarily dangerous expedition, Gertrude received nothing but praise. Perhaps unsurprisingly, however, it made little difference to her relationship with Doughty-Wylie; all that remained, she wrote, was a sense of dust and ashes in her hands. The world around them was changing: a few months after her return from Hayyil, the nations of Europe and beyond were plunged into the Great War. Onto the beaches of Gallipoli in the Dardanelles Strait went Dick Doughty-Wylie and thousands of other British troops on a near-impossible mission to seize the Turkish-occupied coast. Gertrude was at a luncheon party in London when she overheard that he had been killed.

Gertrude had always gone out into the deserts of the Middle East to escape from the world, so it was slightly ironic that the world came to her for advice on the Middle East and its people. Thanks to her appetite for knowledge and her detailed journals, she possessed an almost unrivalled understanding of the complex feuds and allegiances that underpinned the region, as well as the leaders who fuelled them. For the rest of her life, during the war and beyond, she fulfilled a number of high-ranking roles in British diplomacy and intelligence, with specific emphasis on Arab countries. She assisted in forming the new kingdom of Iraq (under British administration) and helped install King Faisal I onto its throne.

[1] Gertrude Lowthian Bell, *The Desert and the Sown* (1907)

RUSSIA
(KYRGYZTAN)
AFGHANISTAN
(TAJIKISTAN)
KASHMIR
CHINA
TIBET
KASHGAR
YARKAND
TASHKURGAN
GILGIT
KHOTAN
Dandan-Uiliq
Tarim Basin
Taklamakan Desert
Kunlun Mountains
75°
80°
40°
N
100 km
RUSSIA
KAZAKHSTAN
MONGOLIA
IRAN
INDIA
CHINA

Sir Aurel Stein

o

On a cold evening in November 1900, Marc Aurel Stein pitched his tent on a high vantage point in the foothills of the Kunlun mountains. He was tired, thirsty and hungry, but there would be no food and very little water until the rest of his men arrived with their ponies, which might not be until the next morning. This didn't worry him greatly; he would quite cheerfully go without his dinner because he was so excited about tomorrow.

At least he could make some tea. He poured the last cupful of water out of his water-bottle and set it over a small stove. It had been an exhausting day, toiling up and down the steep sides of ridges whose dry, stony soil supported few plants and hardly any wildlife. He wanted nothing more than to go to sleep, but the sunset had been so spectacular that he had no choice but to stand and watch as the sea of distant peaks faded from brilliant rose to red and then rich purple. Now, as night fell, another light called him back outside.

The full moon was rising, shimmering at first on the dust haze that cloaked the flat landscape to the north, and then steadily growing in brilliance until Stein could scarcely believe his eyes. Was he really on the edge of the notorious Taklamakan Desert, where no life was said to exist, whose very name signified a place of no return? Maybe the moonlight was playing tricks on him. 'It seemed,' he wrote, 'as if I were looking at the lights of a vast city lying below me in the endless plains.' [1]

In his homeland of Hungary, Stein's boyhood fascination for the exploits of Alexander the Great had fuelled his rise through some of the most illustrious academic institutions of Europe. After studying philology and Indian palaeography in Vienna, Leipzig and Tübingen, he moved to Britain where he focused on collections held in the Ashmolean Museum, the Bodleian Library, and Trinity College, Cambridge. In London, Stein made new acquaintances at the Royal Asiatic Society and the India Office Library, which in turn led to the opening of career prospects in India. In 1887, he accepted a dual post as Registrar at the Punjab University and Principal of the Oriental College in Lahore.

Stein was just 25 when he first arrived in India, but such was his expertise that within a few years he was cataloguing a previously unexamined collection of Sanskrit manuscripts in the library of the Raghunatha Temple in Kashmir. It was an exciting time to be a specialist in ancient languages, because hints of little-known civilisations were coming to light through antiquities acquired by travellers and collectors. These precious relics, sometimes fragmentary, gave scholars a tantalising glimpse into the spread of early Indian culture and religion across the Himalayas and into China.

But Stein suspected that some of these new 'discoveries' - for example, purportedly ancient writings on birch-bark from Chinese Turkestan - might be forgeries. In order to discover the truth, he realised that he must visit the places where these relics were said to have been found. Locating them, however, was never going to be an easy task. The writings of Hiuen Tsang (Xuanzang), a seventh-century Chinese Buddhist monk, offered valuable clues, and Stein memorised his descriptions of the topography of the Tarim Basin. Explaining his focus on Khotan, Stein wrote that this region held particular promise, being located on an important route from China to the Oxus Valley and hence to India. 'Fortunately for our researches,' he wrote, 'archaeology could here rely on the help of a very effective ally - the moving sand of the desert which preserves what it buries.' [1]

It took Stein nearly six months to reach the southern fringe of the Taklamakan Desert, travelling in a slow-moving expedition from Kashmir through Gilgit, Tashkurgan, Kashgar and Yarkand. En route, he heard accounts of a known forger who was manufacturing 'old' books and selling them to European collectors. Near Khotan, a man offered to sell him a birch-bark manuscript; Stein applied an experimental spot of water and the ink dissolved, betraying its recent origin. He was dismayed to think of the months of work already undertaken by his academic colleagues in deciphering identical-looking 'relics'.

At the same time, there was a very real chance of discovering something significant. Stein consulted Badruddin Khan, who was in charge of Afghan merchants in Khotan and could help locate little-known sites in the desert; and he spoke with the Amban or Chinese high official of Khotan, a gentle, elderly man who understood the work that Stein was proposing to undertake. Within a short time, details came in about a place that had yielded promising artefacts, some four days' march from Khotan.

Stein was accompanied by his trusted surveyor, Ram Singh, and his cook, Jasvant Singh, and he hired about 30 local people as guides and workmen. The ponies which had helped him over the high mountain ridges were exchanged for camels and donkeys, loaded with enough food to last several weeks. On 12th December, they were ready to depart.

Immediately, Stein found that marching through drift sand was tough and tiring work. The feet of men and animals sank deeply at every step, and for this reason the camels' speed was reduced to about one mile an hour. Sporadic outcrops of tamarisk bushes provided firewood, and wells had to be dug for water, which often turned out to be bitter and barely potable. Winter was already beginning to bite: water was collected before it froze and then stored in iron containers as ice.

During the daytime, Stein revelled in the 'delightfully pure air of the desert and its repose which nothing living disturbs.' [1] At night, however, temperatures sank to -23°C (-10°F), and he found that his tent, despite its additional serge lining, was 'a terribly cold abode.' [1] He carried a small stove, fuelled by small compressed 'cakes' steeped in paraffin, but its output was not enough for Stein to discard his heavy fur-lined overcoat and boots. Too cold to read or write, he retired to bed, covering his head and face with a fur-lined Balaclava topped with a knitted Shetland hat. His little terrier, Dash, who went with him everywhere, nestled into Stein's bed, snug in his own custom-made fur coat from Kashmir.

On their fourth evening in the desert, the expedition came across some dead trees still rooted in the heavy sand. Although they were shrivelled and bleached, Stein's guides recognised them as poplar and willow, a sign that they had reached a place of ancient habitation. The following morning, Stein found himself standing in the remains of an ancient city, once a flourishing settlement on the Silk Road. He already knew its name: Dandan Uiliq.

Four years earlier, a Swedish explorer named Sven Hedin had come here and guessed at its significance. Hedin was an experienced topographer, but he was not an archaeologist. Occasionally, other travellers from Khotan and further afield had chanced across half-sunken buildings and scraped around in search of 'treasure'. But since its abandonment over a thousand years ago, Stein was the first visitor to Dandan Uiliq who was capable of recognising and recording its history.

Frescoes on a structure that was partly free of sand told him immediately what he was looking at. Representations of Buddhas and Bodhisattvas indicated a place of worship, and from their style he placed it around the eighth century CE, a theory supported almost immediately by some scattered copper coins. He was glad to see that the hunters had not dug extensively, or with any concerted effort. Some damage had occurred, but most of the city, just as Stein had foreseen, was still perfectly preserved beneath the desert sand.

Most of the city was still perfectly preserved beneath the desert sand.

Stein was meticulous in his approach, surveying the site and making careful notes of the depths at which structures and objects were uncovered. Over the next two weeks, he and his men excavated 14 houses and temples. The first of these, a small square shrine, was enclosed by an outer quadrangular passage that might have been used for processions. Inside the shrine, only the feet remained of a once-colossal statue of a Buddha; still propped against it were small painted wooden panels, likely left there by pilgrims. The walls bore faded Buddhist frescoes, as well as a painting of some riders on horses or camels holding an offering to a swooping bird. Stein even unearthed brooms made of grass stalks, evidence that great care had once been taken to keep these sacred sites free from the encroaching sand.

Within days, fragments and bundles of manuscripts emerged from the loose sand, some of them pierced with holes through which a cord was once threaded. Each bore the same curious symbols. Stein recognised it as the Brahmi script of the Gupta period, named after an ancient Indian empire that existed from the fourth to the sixth centuries CE. From the religious nature of their contents, Stein believed that they had once belonged to the library of a monastic community; they were later identified as one of the oldest extensive copies of the Vajracchedikā or Diamond Sutra of the Mahayana School of Buddhism.

Another manuscript, found elsewhere, had a more mundane purpose: a petition for the return of a donkey which had been loaned to two people and was still missing after 10 months. Aside from its intriguing story, this paper was of great value because it was precisely dated to the Ta-li period, corresponding to 781 CE, and it also revealed the older name of Dandan Uiliq, which was Li-sieh.

Meanwhile, old stories were preserved in wall paintings. A crowned figure with a rat's head was identified as the King of the Rats who, in Buddhist legend, saved Khotan from invasion by inciting thousands of his followers to chew through the harnesses of the aggressors' horses. Another panel showed a Chinese princess who was sent to marry a prince of Khotan, and smuggled silkworm eggs and mulberry seeds out of China by putting them in her hair. The secret of China's silk production had always been jealously guarded, and it was said that the mulberry groves of Khotan became one of the earliest and richest sources of silk outside mainland China.

Why had Dandan Uiliq been abandoned? Sven Hedin had concluded that the Keriya river, a major source of water for drinking and irrigation, had changed its course and forced the city's inhabitants to leave. Stein, who had gained a more precise idea of the city's timeline, pointed to the threat of invasion from Tibet. He also suggested a reason why it was never reoccupied by later generations: a milder climate had led to the gradual shrinking of glaciers in the nearby mountains, leading to a reduction in river volumes.

It was a surreal experience to see houses emerging with so many signs of human life still intact. The all-encompassing silence, together with the sand-blasted and bleached tree trunks, gave Dandan Uiliq an eerie atmosphere which unnerved some of Stein's team. Stein himself was not given to flights of imagination, but he did suffer a few moments of panic when he lost his bearings at twilight on Christmas Day, having walked a mile or two away from the site to inspect some dunes. He had left his compass behind, and was greatly relieved to hear the shouts of his friends who had gone looking for him.

Living in the Taklamakan Desert was demanding at the best of times, but Stein was also in pain from a troublesome tooth which a barber in Khotan had tried - and failed - to extract. Self-medicating was the only option, and at night he laboriously melted a small quantity of ice over a candle so that he could dilute a few drops of chlorodyne, a painkiller that contained laudanum.

But nothing clouded Stein's delight in his own extraordinary circumstances. As he fed Dash with some tasty morsels from his Christmas dinner, he allowed himself to reflect on his own happiness and good fortune. 'I sometimes doubt,' he wrote, 'whether even the friends whose kind thoughts turned towards me that evening from the distant South and West could realise how cheerful is the recollection of the Christmas spent in the solitude and cold of the desert.' [1]

'How cheerful is the recollection of the Christmas spent in the solitude and cold of the desert.'

Stein left Dandan Uiliq on 4th January 1901. During his three-week encampment there, he had become 'almost fond' of the stark scenery, and he was looking forward to inspecting further sites before returning to modern civilisation. At the end of

April he was back in Khotan, regretfully browsing the bazaars in search of gifts for friends and family. Then he retraced his steps through Yarkand and Kashgar, with a collection of finds packed in 12 large boxes and countless more described in reports, illustrations and photographs.

Even in Stein's own perpetually modest estimation, this first expedition had been a success, and he longed to set out on another. But since he had made so many significant discoveries, his first priority was to present them to the public, and he began writing a two-volume work entitled *Ancient Khotan*. 'Incessant desk-work,' he complained, 'kept me imprisoned in my little cottage above the Dal lake near Srinagar from morning to dusk.' [2] Before organising another expedition, he also had to obtain leave from his 'exacting official duties', in other words his new day job as Inspector of Schools in the Punjab.

A few years later, he was back in the Taklamakan Desert and contemplating the excavation of another buried site. From beneath his feet, a sea of yellow dunes rolled away into the distance, punctuated here and there by splintered posts that resembled the ribs of a decaying ship. The desert had utterly bewitched him, and he wanted to immerse himself in it completely. He wrote: 'For the first few days I found it difficult to confine my thoughts to the multifarious tasks which claimed me, and not to listen inwardly to the Sirens' call from the desert northward. How joyful it seemed, to be free for pushing farther and farther into that unknown waste beyond the last dunes on the horizon!' [2]

In all, between 1900 and 1930 Stein undertook four major expeditions, tracing ancient caravan routes along the Silk Roads through Central Asia and exploring sites linked with Alexander the Great. He found the first material evidence that China had used silk as a writing material before the invention of paper, and he was the first archaeologist to visit the Caves of the Thousand Buddhas near Dunhuang. A mail bag from the fourth century CE, discovered near a desert watch-tower, contained five letters in the 'lost' language of Sogdian. More prosaically, at Loulan he stumbled across a tape measure that had been dropped by Sven Hedin, which he returned to him at a dinner hosted by the Royal Geographical Society in London.

Focused as he was on physical structures and artefacts, Stein also experienced sudden glimpses of less tangible things, and he never forgot the moonlit vision of a great city that appeared to him over the Taklamakan Desert. At times, it was as if he could sense the tread of human feet through these vast, barren spaces. In a paper published in the *Scottish Geographical Magazine*, he wrote: 'I might talk for hours about the strange observations and experiences which... made me forget as it were the lapse of long ages. More than once as I sat in the evening amidst the debris of some commanding watch-station the slanting rays of the sun showed up quite clearly a curiously straight, furrow-like line keeping parallel to the wall and about twenty feet within it wherever there was a well-preserved stretch of it. Repeated examination proved that it was a shallow but well-defined track worn into the fine gravel soil by the patrols and others who had tramped along here for centuries.' [3]

[1] M Aurel Stein, *The Sand-buried Ruins of Khotan* (1903)

[2] M Aurel Stein, *Ruins of Desert Cathay* (1912)

[3] Dr M Aurel Stein (1910) 'Explorations in Central Asia 1906-08', *Scottish Geographical Magazine*

TURKMENISTAN
Meymaneh
Paropamisus Mountains
HERAT
Obey
Chist--e-Sharif
Minaret of Jam
Chaghcharan
Yakawlang
BAMIYAN
Maidan Shahr
KABUL
AFGHANISTAN
PAKISTAN
N
600 km
62°
64°
66°
68°
36°
34°

Rory Stewart

Rory Stewart admits that he has always had difficulty explaining why he wanted to walk across Afghanistan. Geographically, at least, it made sense: he had already spent 16 months walking across Iran, Pakistan, India and Nepal, but in December 2000 he'd had to leave out Afghanistan altogether because the Taliban government refused him entry.

Then, in December 2001, he heard that the Taliban had fallen. Here, surely, was his chance to fill in the gap. From Nepal, he drove back to Afghanistan with the intention of walking from Herat to Kabul, a distance of about 497 miles (800 km). The new US-led coalition government had only been in place for two weeks, meaning that tensions would still be high. In the desperately poor communities through which he'd be passing, the arrival of an outsider was likely to provoke curiosity at best, and hostility at worst.

Rory knew that he was about to take a considerable risk, but he also believed that he could survive. He had served in the British Army and in the diplomatic service. Formidably well-informed about foreign affairs, he spoke conversational Dari. Of the two available routes from Herat to Kabul, he chose the shorter, more direct one, because he had heard that the Taliban were still fighting along the southern route. The only problem was that his intended path ran through the mountains, which in January would be under deep snow.

In the Foreign Minister's office in Herat, Rory explained who he was, and produced papers confirming his purpose: he was a Scot, interested in the environment and medieval history, travelling alone and in peace. The meeting was relaxed and Rory was confident of his prospects. Then, in his hotel, he was met by members of the Security Service and forcibly escorted back to the Ministry. In an atmosphere of heightened suspicion, he learned that two armed escorts would accompany him all the way.

Rory was concerned that the presence of such men would endanger him in places where he hoped to seek hospitality, but he was told that if he refused he would be expelled. His companions, two soldiers, were introduced. Rory hoped that they would get tired or bored and abandon him at an early stage. But just eight kilometres outside Herat, a third soldier, who was a relative of one of his escorts, decided to join them.

Close to one of the traditional junctions on the Silk Road, a US journalist clattered up in a jeep and asked the obvious question: why? Rory replied that he saw a magic in the act of leaving footprints across Asia. He wanted to see where the cultures of Islam and Buddhism merged into one another, in this place between the deserts and the Himalayas. But, said the journalist, surely he could see how dangerous it was? Rory didn't answer that. Instead, he spoke about the Prussian blue of the sky, the quality of the air, and added that he felt he had been preparing for this journey all his life. The journalist jotted down some notes.

He wanted to see where the cultures of Islam and Buddhism merged into one another.

As the shadows lengthened, so did Rory's stride. This brief interview had unsettled him: would it be published as an obituary? Unlike the countries he had just traversed, the surrounding landscape was almost devoid of feature and offered little in the way of distraction. Sand and gravel stretched away to low hills on either side; poverty and four years of drought meant that signs of human occupation were almost non-existent. Nearby, a landmine suddenly exploded at random, making the ground tremble, and a pall of black smoke rose from a field.

Rory was reluctant to stop at the bazaar of Herat Sha'ede, because he felt he could walk further before nightfall. However, the decision was made for him. The soldiers wanted to take him to a cemetery, and he wondered if he was about to be killed. It crossed his mind that if he was, people would think he had been foolhardy. But the soldiers merely wanted to show him things of interest. The cemetery had a shrine and a mausoleum, and next to it was a garden of roses and green, freshly mown grass. The peace and beauty of it took him by surprise. They stayed there a while and then, as the setting sun sank through layers of orange dust, one of the soldiers picked a flower and tucked it into Rory's cap.

A couple of days later, they were seeking overnight shelter in a caravanserai, one of the traditional roadside stopping-places along the Silk Road. Rory thought about the precious trade goods that this mud-walled building might once have housed: lapis lazuli from Afghanistan, amber from the Baltic and bound for Tibet, Buddhist texts on birch-bark scrolls. But as they approached, one of his fears was realised. Two young boys emerged, obviously sent out from the building on purpose, and told them that no one was at home. The soldiers levelled their rifles at the children. Maybe it was intended as a gesture. Appalled, Rory laid his hand on one of the weapons and said, 'No.'

In the next village, nervous inhabitants were persuaded to offer hospitality at gunpoint. Listening to the conversation in a communal room, Rory observed that their host never took his eyes off the weapons. He felt excruciatingly uncomfortable. As his companions swapped stories of bravery, he spread his sleeping bag in a corner and tried to go to sleep.

Fields of land mines and armoured personnel carriers, now abandoned, marked their onward route. The soldiers pointed out sites of recent ambushes and violence. As they got further away from Herat, the influence of the new government diminished: a party of armed men, loyal to a local military commander, detained and questioned Rory's companions. On their release, Rory insisted that they all walk on into the night, enduring cold rain that rapidly turned to snow.

A quarter of the way through his journey, Rory learned that the soldiers would soon be leaving him. Although he had come to a better understanding of them, it was a huge relief. Maybe he had walked further and faster than they expected; they had blisters on their feet, and ahead lay the Paromamisus mountains. He was glad to regain the self-reliance of a solo traveller.

But he was not quite alone. In the hill village of Dahan-e-Razak, his host asked him if he would like a dog. A great mastiff, golden-haired and yellow-eyed, was lounging on a blanket. He had spent his life guarding flocks of sheep: the villagers assured Rory that he had killed many wolves, and his embattled appearance was the result of many dog fights. He was certainly not a pet. He was being offered to Rory as a protector against hostile tribes and wolves. Rory thought he seemed old and tired, but the stump of his tail moved slightly in response to a greeting. When Rory headed up into the hills later that morning, the dog was by his side.

He called him Babur. Unknowingly, at least to begin with, Rory had chosen to follow the route taken by another Babur, who was the prince of a remote kingdom in Uzbekistan. In the early 1500s, Babur lost his lands in Central Asia; with a small band of followers he fought to regain power, crossing from Herat to Kabul in winter before pressing east towards Delhi where he founded the Mughal Empire. Rory had begun to read Babur's diaries and found him an honest and careful observer, describing his experiences without embellishment. Babur's canine namesake, perhaps driven by a similar urge for conquest, paused every few paces to lift his leg and mark his new territory. Rory wondered how long, at this rate, it would take him to cover the remaining 373 miles (600 km).

And then he saw something that forced him to stop in his tracks. High in the hills, where the Hari Rud river flows through a narrow gorge, stands the Minaret of Jam. From an octagonal base only nine metres in diameter, this free-standing finger of exquisitely adorned brickwork rises 213 feet into the sky. Around its slender column, Kufic lettering spells out texts from the Koran. A circlet of blue tiles, still as vivid as the day they were made, preserves the name of Ghiyath al-Din Muhammad, a Sultan of the Ghorid Empire.

In the near-darkness of its interior, Rory began to clamber up a spiral staircase, but missed his footing on the crumbling stonework and came crashing down, knocking his head and scraping his hands. The 800-year-old tower seemed to shake slightly as he fell, and for a horrible second he wondered if he would go down in history as the man who died while accidentally demolishing the last surviving masterpiece of Ghorid architecture. Dusting himself off, he tried again, and was rewarded with breathtaking views from the lantern at the top. It was then that he noticed a line of excavation trenches on a far slope.

Back at ground level, a newly-appointed official, whose job it was to protect the tower, explained what was happening. He brought out a tray of finds: bronze dice, fragments of delicate porcelain, pieces of carved marble, bone beads. All had been dug up by local people, and would soon be on their way to international dealers. As an unforeseen by-product of the newly-installed government, artefacts that could help reveal the history of this place were leaving the country before they could be properly examined.

Before the Russian invasion in 1979, a handful of visiting archaeologists had speculated that this was Firuzkuh, also known as Turquoise Mountain, the summer capital of the Ghorid Empire. A few excavations had been made, but no evidence was found. Now, the ground was giving up its treasures. As Rory clambered over the rough trenches, he saw that the villagers had uncovered what looked like an ancient city. He speculated that the minaret may have been one of a pair that stood on either side of a huge Friday Mosque at the bottom of the gorge. When Genghis Khan or his successor razed the city in the 13th century, this minaret was spared,

He memorised his own verbal route-map, which he thought of as 'a song-of-the-places-in-between'.

perhaps to prove the magnificence of what had been obliterated.

Rory did not carry a map with him, in case he was accused of being a spy. But along his path, he discovered that villagers could chant a list of all the settlements that lay ahead, coupled with the names of their respective headmen. He memorised his own verbal route-map, which he thought of as 'a song-of-the-places-in-between'. It served not just as a way-finder, but as useful proof of his acceptance at previous stopping points. Villagers were naturally hospitable to travellers, but their experience of the war still made them wary. Words that echoed their own traditions spoke to them more clearly than any official papers.

Nevertheless, such acceptance wasn't always guaranteed. Once, Rory was shot at and had to run for cover. At the nearest inn, he came face-to-face with the gunman, a local military commander, who now acknowledged him with a show of courtesy and even wrote a letter of introduction to show future hosts. Babur's presence also attracted attention, not least from packs of half-wild village dogs that Rory had to beat off with a stick. A horseman enquired if Rory was looking to enter him in an organised dogfight; Rory explained that he was not.

As they walked into the teeth of winter, Rory and Babur followed high, narrow paths between walls of snow; sometimes, the easiest track to follow was along a frozen river, but when Rory slipped through the ice he had to haul himself out and struggle on against a bitter wind. Over the Zarin Pass, where two villagers had insisted on accompanying him for safety, the snow came down thick and fast. Rory's guides saw him into the Zarin Valley, bound for Yakawlang, and then turned to retrace their steps.

At a bazaar in the province of Wardak, Rory's customary greeting in Dari was not returned. Sensing danger, he kept walking. Questions were fired at him. Where was he from? Where was his passport? Who did he support, America or the Taliban? Rory replied that he was an Indonesian professor of history and had no opinion on the matter. It was an uncontroversial identity, one that they found hard to dispute. Armed men followed him suspiciously; one was introduced as the Taliban head of the village. A suggestion was made: why didn't Rory go and look at the river? There was laughter when he refused, and he demanded to know why. Because, they said, he would have been killed if he'd complied. A convoy of pick-up trucks distracted them. He walked on.

The military barracks at Maidan Shahr, on the outskirts of Kabul, was occupied by soldiers serving the new interim government. Rory's thin and dishevelled appearance almost fooled them into thinking he was a member of Al-Qaeda. They gave him sweet tea, a hot meal, and shelter for the night. He relaxed in an atmosphere of comradeship. When he stepped out onto the dusty road for the last time, he felt a deep sense of completion, not just of his journey, but of something within himself.

Looking back at his journey through Afghanistan, Rory says that he gained so much, simply from walking: 'It gave me transformation, and admiration for practical knowledge as opposed to book knowledge: the kind of knowledge that is informal and located in a very particular piece of ground, rather than structural, systematic and theoretical.' [1]

Afghanistan struck him as being startlingly different from the other parts of his journey. He explains: 'Both north Iran and western Pakistan share deep cultural and historical links with Afghanistan, so in some ways the borders are artificial. Nevertheless, I saw things in Afghanistan that I never saw anywhere else. For example, in other countries, people talked about villagers accompanying you from one village to another, but it was only in Afghanistan that I saw this. And only in Afghanistan was the hospitality so extreme and so consistent: people much poorer than myself, often at risk of their lives, would put me up in their houses. I felt deeply privileged to see a culture that was, in those respects, so much more impressive than ours.' [1]

Rory also observes that Afghanistan is a country of tragedy in terms of its development, and most recently in the oppression of women. He says: 'I struggle to balance those negative aspects with the incredible grace, courage, honour and generosity that I witnessed daily.' [1]

Three years after Rory completed his walk, he was invited by the Prince of Wales (now King Charles III) to develop an initiative in traditional Afghan arts and architecture, with an emphasis on saving crafts that were in danger of dying out. Rory moved to Afghanistan in 2005 and stayed for three years to establish the project, which was called Turquoise Mountain. 'Nothing,' he says, 'has been so inspiring in my life.' [1]

Work began on restoring the old city of Kabul, Murad Khani, whose decaying historic buildings were under threat of being demolished. Rory recalls that the district was effectively a slum, seven feet deep in garbage, with no water or electricity. But, he adds, there were 'traces of incredible beauty and civilisation and architecture and gardens, hidden amongst the detritus and the craters from rocket fire.' [1]

Artisans now work in the restored buildings, and their products are reaching international markets. Accredited training is provided in skills ranging from calligraphy and ceramics to weaving and woodwork. Turquoise Mountain has also established a number of schools and health clinics.

'It was such an extraordinary project,' says Rory, 'partly because we started with volunteers who knew very little, and we were improvising as we went along.' [1] He was particularly proud to hand over Turquoise Mountain to his wife, Shoshana: 'It is such a privilege to help give birth to something that takes on organic life that goes far beyond your own conception.' [1]

[1] Rory Stewart, reflections on his journey in Afghanistan shared with Jo Woolf (June 2024)

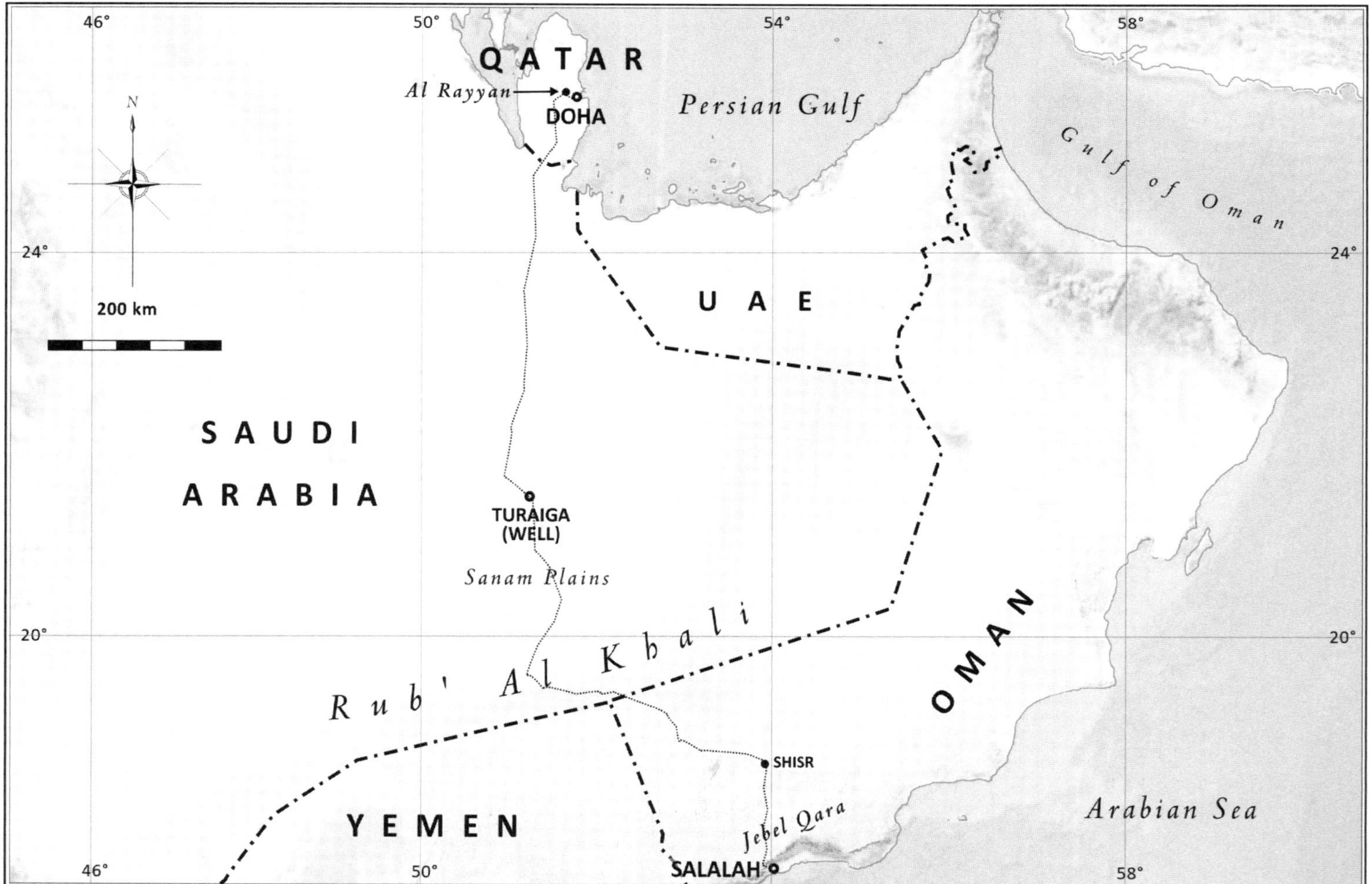
QATAR
Al Rayyan
DOHA
Persian Gulf
Gulf of Oman
UAE
SAUDI ARABIA
200 km
N
TURAIGA (WELL)
Sanam Plains
Rub' Al Khali
OMAN
SHISR
Jebel Qara
SALALAH
YEMEN
Arabian Sea
46°
50°
54°
58°
24°
20°

Mark Evans

On 10th December 1930, a young civil servant named Bertram Thomas and his companion, Sheikh Saleh bin Kalut, took a string of camels and headed north into the hills above Salalah. They were walking into a vast, unmapped region; in effect, they were about to vanish off the face of the Earth. No one knew of their departure, because Thomas had planned his expedition with absolute secrecy. The odds of their survival were slim. But 58 days later they emerged in Doha, to worldwide astonishment and acclaim. Together, Thomas and bin Kalut had crossed the world's last great unexplored desert: the Rub' al Khali, or Empty Quarter, in Southern Arabia.

On 10th December 2015, an expedition comprising three men and four camels departed from Salalah with the aim of retracing Thomas's route. If they succeeded, they would be the first people in 85 years to recreate his journey.

The initiator of this daring expedition was Mark Evans. Living and working in Oman, Mark had read about the great crossings of the Empty Quarter by Harry St John Philby and Wilfred Thesiger, but he'd been surprised to discover that a comparatively little-known man from Bristol had pre-empted both of them. After his triumph, Thomas had fallen into relative obscurity, and Mark felt that his achievement deserved more widespread recognition. He explained: 'I cannot but admire what Thomas and bin Kalut achieved, and if this journey puts them back where they should be, then it will all be worthwhile.' [1]

If Mark's own path to the Middle East were traced on a map of the world, it would begin in Shropshire, wander briefly down to London, and then head abruptly north, to Greenland, Spitsbergen and Arctic Canada. Then it would slide south again to Kenya and east to Bahrain, where, as Head of Geography at St Christopher's School, Mark happened to fall in love with the desert.

Mark's first taste of the wilderness came in northern Norway, on an expedition set up by the British Schools Exploring Society. Aged 17, he spent six weeks living on a glacier, helping with scientific research and sharing a tent with two people he'd never met. The experience opened a door onto a wider world he never knew existed. It was, he remembers, 'a bit like Bilbo Baggins stepping out of the Shire!' [2]

Keen for more Arctic travel, Mark and a friend, Nigel Harling, re-traced Fridtjof Nansen's ski crossing of Greenland. Afterwards, Mark followed the trail of 19th century navigator William Edward Parry in his quest for the Northwest Passage. As an expedition leader for groups of young people, he spent two years living in Svalbard. 'When I reflect about my expeditions,' he says, 'there's always a historical connection... I have this fascination with the people who went before. From these old explorers, I learn a lot that enables me to move forward.' [3]

Mark's arrival in Bahrain was in direct response to his need for a job. He admits that he came to the Arabian Gulf for all the wrong reasons, but he stayed for all the right ones. He bought a LandRover and spent three years tagging along on desert journeys with other travellers, sleeping on the roof-rack and learning about desert navigation and survival.

'I feel completely comfortable out here,' he explains. I feel very uncomfortable in noisy, crowded places. This is where I find contentment.' [3] And having found that contentment, Mark wondered how he could help others find it too. He realised, as he sat around a desert campfire, that the answer was right in front of him: 'The fireplace of the desert... is the oldest university on earth. Around the fire of the desert, people have sat and discussed issues and challenges for thousands of years.' [3]

'The fireplace of the desert is the oldest university on earth.'

Mark's idea led him to establish Outward Bound Oman, the first Outward Bound school in any Arabic-speaking country. Beneath the desert sky, far from the distractions of modern technology, groups of young people from different nations and cultures develop new connections and debate solutions to global issues. Mark says: 'We've used that place of learning as a catalyst to inspire future leaders... our motivation is mission-driven because we believe that we can make a difference to people's lives.' [3]

For Mark, the satisfaction of introducing newcomers to the desert ran parallel to the fascination of explorers who had been there before. Inspired by Thomas's *Arabia Felix*, which tells the story of his Empty Quarter crossing, Mark pored over his black-and-white lecture photographs at the Royal Geographical Society in London. At the Faculty of Asian and Middle Eastern Studies in Cambridge, which houses Thomas's diaries, letters, and even some cine film, Mark was astonished to see that in ten years, only six people had accessed the collection; and at Thomas's family home in Pill, near Bristol, he had to search hard to find both his birthplace and his grave.

Thomas had mapped his route in detail, marking waterholes, tribal territories and topographical features. However, in the intervening decades, political borders had been thrown up across his path, discouraging any attempt to retrace it. Mark's ambition had already set seed in his mind, but he would be crossing from Oman to Saudi Arabia and then to Qatar, and in the desert their respective borders were unmarked by any checkpoints, creating logistical difficulties. But fortune lent a hand, and he found his plan supported by influential Patrons: His Highness Sayyid Haitham bin Tariq Al Said, Oman's Minister of Culture and Heritage; Sheikh Joaan bin Hamed Al Thani, Minister of Culture and Heritage in Qatar; and HRH The Prince of Wales (now King Charles III).

Occupying some 650,000 square km (251,000 square miles), the Empty Quarter is one of the hottest, driest and most inhospitable regions on the planet. To complete his 1,300-kilometre (808-mile) trek, Thomas needed to negotiate safe passage with a succession of tribes whom he relied on to guide him from one waterhole to the next. Their nomadic way of life has all but disappeared, so Mark focused on replicating that spirit of trust and cooperation in his own team. His companions were Amur Al Wahaibi, a Bedouin guide and an expert in desert survival, and Mohamed Al Zadjali, a senior instructor with Outward Bound Oman.

After leaving Salalah, the team climbed up into the Jebel Qara mountains whose slopes were forested with frankincense trees, and crossed a network of steep-sided wadis or dry river beds. The desert itself was ten days' walk from Salalah, and they got into a routine of rising at dawn, walking for five hours, resting in the midday heat, and walking again until evening. After a simple meal by a campfire, they lay down to sleep beneath a sky that was alive with shooting stars.

Almost every day, while on the fringes of the desert, the team were greeted and hosted by well-wishers, some of whom had parents or grandparents who had travelled with either Bertram Thomas or Wilfred Thesiger. At the small desert community of Shisr, the reputed location of the ancient city of Ubar, a feast and a joyous welcome awaited them; this generous hospitality was a welcome boost to their spirits before they left the stony plains and entered the dunes.

Referring regularly to Thomas's diaries, Mark appreciated the resilience of earlier explorers. The team were already nursing blisters and cracked lips; Mohamed had severe pain in his feet from walking on loose stones, and discarded his shoes gratefully - as they all did - on reaching soft sand, preferring to walk in socks with double-woven soles that had been knitted by Bedouin people.

Amur, an experienced camel handler, had chosen four animals from the Royal Camel Cavalry of Oman. Today's camels, however, are nowhere near as hardy as those of Thomas's day, and so for a few weeks they had attended a kind of 'camel boot camp' about 200 kilometres north of Salalah, in order to get fit enough for the journey. Because camels are unable to walk for long distances on hard ground, Mark, Amur and Mohamed met up with them about twelve days after their departure, on the southern edge of the sands. Then, quite quickly, Mark and Mohamed had to learn to ride them. Camels can cover between five and six kilometres per hour, and for the expedition that represented a huge saving in time and energy.

That, at least, was the theory. Despite learning from Amur the commands for 'stop' and 'walk on', Mark and Mohamed struggled to stay aloft on their lively camels. One of the four was a challenge even for Amur to control, and since she was inciting the others to rebellion she was eventually deported back to civilisation in the back of a truck. 'It's very difficult to ride them,' observed Mohamed. 'There is too much to learn about this beautiful animal.' [1]

Everyone had the excruciating discomfort of sand grains in their eyes, nose, ears and mouth.

Entering the heart of the Empty Quarter, the team trudged up and down a succession of steep dunes with their reluctant camels in tow. Reading Thomas's account from 1930, Mark was amazed to see that his predecessor physically scooped a path through the dunes for his animals to pass through. Easier terrain lay beyond, but then the wind shifted to the north and started to whip up the sand. The sky darkened, making navigation difficult, and soon everyone had the excruciating discomfort of sand grains in their eyes, nose, ears and mouth. When the storm settled, a more successful bid was made to ride the camels, and a quest began for the nearest waterhole.

Because many of the waterholes that were used by Thomas have now been lost, Mark and his team could not rely on natural sources of water alone. Nor could they carry all that they needed to survive. A camel needs to drink every three days, and will consume about 20 litres in seconds. Meeting them at regular stages in 4x4 vehicles were expedition support members John Smith and Sim Davies, who kept the team supplied with water, as well as food for the men and fodder for the camels.

In between, waterholes still had to be dug out and the water tasted for quality; on finding a good supply, Mark could readily imagine Thomas's relief on reaching a life-giving well. Breakfast on Christmas Day consisted of baked beans, a tribute to Thomas who enjoyed the same treat and described it as 'festive fare' after a strenuous march.

On crossing the border into Saudi Arabia, they were met by Sheikh Mubarak bin Kalut, great-grandson of Sheikh Saleh, bringing with him the beautifully crafted *khanjar* or short curving knife which his ancestor had carried on his trek over the sands. Honouring the spirit of his great-grandfather, Mubarak guided them through the dunes. His whole family, he said, were overjoyed that this journey was being made again, reviving the glories of the past.

Deep in the dunes of Saudi Arabia, with medical facilities hundreds of kilometres away in any direction, some worrying health issues developed. Mohamed suffered from a persistent pain in his abdomen, so the expedition rested for a day while he recuperated and drank lots of fluids; and Mark's eyes reacted badly to the intense sunlight, streaming so profusely that he couldn't read his compass. A sick camel also caused concern, as she was refusing to eat. Amur suspected that it was caused by a broken tooth. Luckily, these issues resolved themselves and the trek continued.

There was cause for optimism, too, in the good progress that Mark and Mohamed were making with their camels. It was by no means the most comfortable form of transport, and the animals were prone to take off suddenly and without warning, forcing their riders to cling on for dear life; but there was a sense of greater understanding and confidence. All in all, some serious obstacles were being overcome, and in the evenings, with the aroma of Amur's desert bread baking over a fire, the team relaxed in the extraordinary peace of their surroundings.

Almost imperceptibly, the desert was casting its spell. One night, Mohamed woke up and was astonished to see that the sky was white. 'Thousands and millions of stars,' he said. 'I really love the place.' [1] Away from his cellphone and the internet, he was finding ample time just to think. Likewise, Mark was renewing his own passion for the desert: 'No footprints, no rubbish, no light pollution, sleeping on the sand every night, shivering

'No footprints, no rubbish, no light pollution, sleeping on the sand every night, shivering mornings, sweltering days - absolute heaven.'

mornings, sweltering days - absolute heaven.' [3]

However, there were still hazards to negotiate, and their destination of Doha lay 400 km (249 miles) away. One morning they awoke to strong winds driving black clouds that soon started to empty themselves of water. Within minutes, their clothes were soaked and the camels were bellowing in alarm; the panicking animals were led to graze on a clump of bushes until the storm passed. Further on, under Amur's expert eye, an area of sabkha or marshy quicksand was crossed with extreme care. If the camels' legs had sunk into it, they would have been unable to get out.

The highest point of their trek had been an altitude of 800 metres (2,625 ft) in the mountains just north of Salalah. Now, as they began to make the long descent towards the sea, fog and heavy dews chilled them as they walked, and night-time temperatures dropped to 5°C (41°F) The camels, however, enjoyed browsing on the more abundant vegetation. Foxes, dragonflies, bees and butterflies were seen, and every morning the movements of nocturnal wildlife were preserved by fresh tracks in the sand.

Collaborating with the Green Arabia Project, the team examined evidence of early human presence in the desert, in the form of scattered stone arrow-heads and curious rock-face inscriptions. Today, the Empty Quarter seems almost devoid of life, but 10,000 years ago the wadis were filled with water and the landscape was much more fertile. The aim of the Green Arabia Project was to understand more about the desert's history of habitation, with a view also to predicting its future.

Just over the border in Qatar, the team were welcomed by NOMAS, an organisation dedicated to reconnecting young Qataris with their desert heritage. Fresh camels had been sent by the Emir of Qatar, Sheikh Tamim bin Hamad Al Thani, in case their own animals were exhausted. Here, they began to walk back into civilisation; after six weeks of travelling on rock and sand, a tarmac road seemed strangely unfamiliar. On 27th January, 49 days after they had set out, they entered Doha.

As the walkers approached the old fort of Al Rayyan that had marked the end of Thomas's journey, there was a sense of history coming full circle. The expedition's Qatari Patron, Sheikh Joaan bin Hamad Al Thani, was there to greet them, along with the Ambassadors of the United Kingdom, Saudi Arabia and Oman. Reporters and photographers relayed news of their arrival to a global audience: a far cry from Thomas's day, when he had to sail by dhow to Muharraq in Bahrain in order to telegraph news of his crossing to the outside world.

Among the team, there was a quiet sense of pride. The desert had forged a special bond between them. Mark paid tribute to his two companions: 'When you are totally exhausted at the end of yet another 30-kilometre march, it is so easy to say, "Not tomorrow - let's have a rest." Those words have not been uttered once, and these two individuals are wonderful role models for young people in Oman.' [4] Mohamed added: 'I am hoping to send a message to inspire Omani youth - and the message is that Omanis are always able to endure difficulties. Not just endure them, but overcome them, always.' [1]

Mark also expressed profound admiration for Bertram Thomas and Sheikh Saleh bin Kalut. 'To walk into nothing like they did - there was no map to follow, and your job was to probe beyond that and see if there was a way through - these people were quite extraordinary. We just demonstrated, on our own 49-day challenge, that there are times when you want to give up. But positive thinking - having a plan, sticking to it, and working together - will yield success.' [1]

Between them, Mark, Mohamed and Amur had lost 24 kilos in weight, but they had gained something much less tangible. Amur found words that spoke for them all. 'When I come to the desert,' he said, 'I feel serenity in my heart and soul.' [1]

o

> Since his trek across the Empty Quarter, Mark has followed in the footsteps of another explorer, Harry St John Philby, who crossed the same desert from east to west, and in January 2025 he retraced another of Bertram Thomas's journeys along the coast of Arabia. He says: 'A theme to my journeys increasingly is trying to show another side to Arabia, beyond the headlines that tend to dominate the news, because the people here are the kindest people you'll ever meet.' [3]

o

[1] *Crossing the Empty Quarter*, documentary produced, directed and written by Simon Gallimore (2017)

[2] Mark Evans, talk to Yorkshire Dales Millennium Trust (2021) https://www.youtube.com/watch?v=yd05vOgqlfk

[3] Mark Evans, *Heart of Arabia* podcast (2023)

[4] Mark Evans, *Crossing the Empty Quarter* (2016)

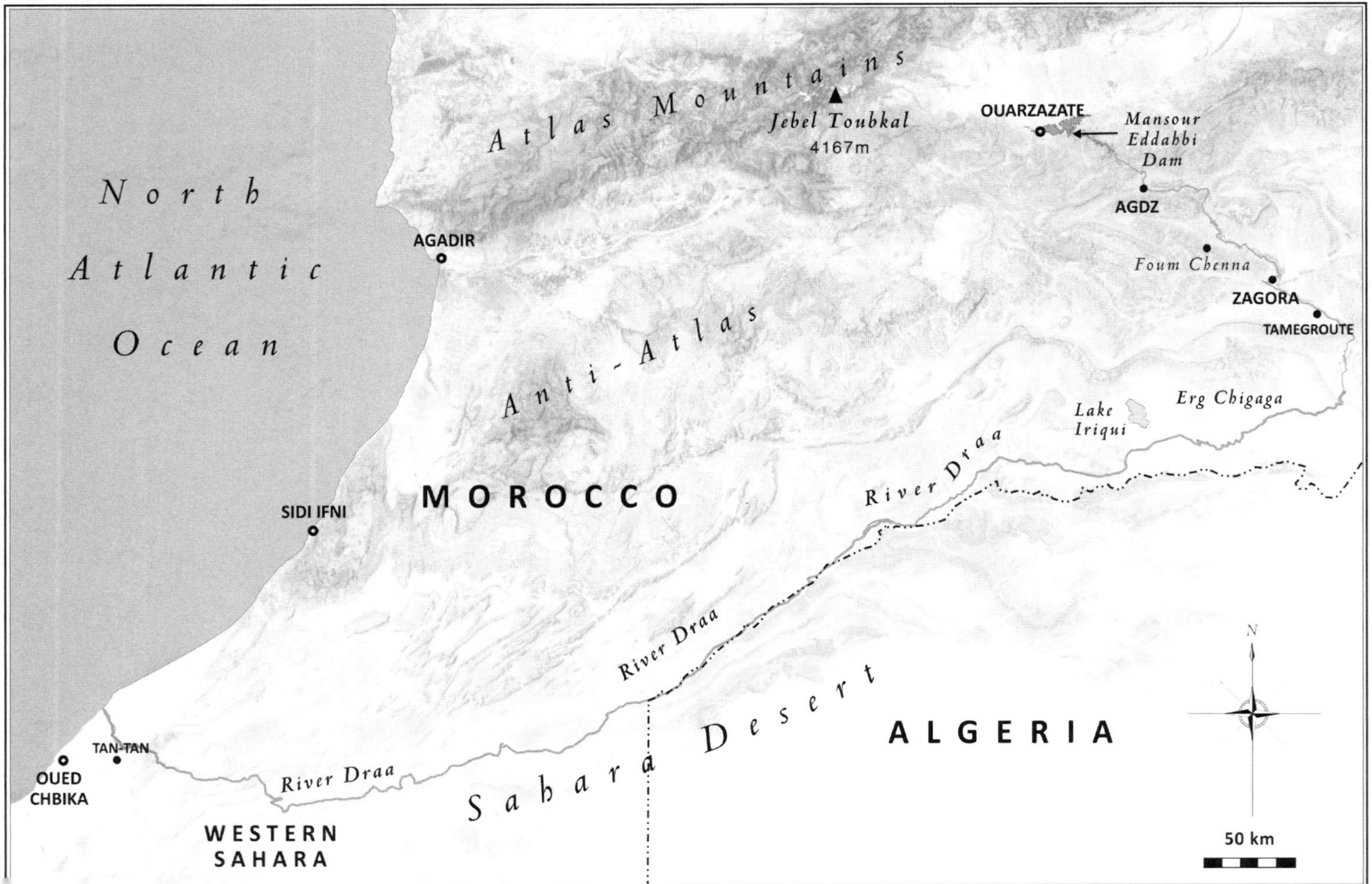

Atlas Mountains
Jebel Toubkal
4167m
OUARZAZATE
Mansour
Eddabbi
Dam
AGDZ
Foum Chenna
ZAGORA
TAMEGROUTE
North
Atlantic
Ocean
AGADIR
Anti-Atlas
Erg Chigaga
Lake
Iriqui
MOROCCO
SIDI IFNI
River Draa
River Draa
River Draa
N
Sahara Desert
ALGERIA
TAN-TAN
OUED
CHBIKA
WESTERN
SAHARA
50 km

Alice Morrison

'Maybe the definition of complete happiness,' says Alice Morrison, 'is when you see the oasis at the end of your day's walk and you know that's where you're going to pitch your tents and spend the night. You've got a fire, you've got food, you've got water - you've got everything to look forward to.' [1]

It's January 2019, and Alice is making a short video for her YouTube channel. She is on an expedition that aims to trace the course of Morocco's longest river, the Draa, from its headwaters in the Atlas Mountains to its entry into the Atlantic Ocean. On a walking journey of some 1,200 kilometres (746 miles) she will travel through an arid, stony landscape punctuated by oases and palm groves, and she will plunge into vast fields of sand dunes on the edge of the Sahara. One of her aims is to see how the natural water sources of this region have been affected by human intervention; she also wants to get to know some of the diverse peoples of Morocco. At the deepest level, she is simply responding to a strong urge to explore.

'You've got a fire, you've got food, you've got water - you've got everything to look forward to.'

Alice was born in Edinburgh, but her childhood was spent roaming freely in the African bush and the foothills of the Mountains of the Moon in Uganda. She studied Arabic and Turkish at Edinburgh University, taught for two years in Cairo, and then embarked on a career in media and broadcasting that took her to Dubai and back to the UK. In 2014 a friend persuaded her to enter the gruelling Marathon des Sables in the Sahara Desert. In order to train, she moved to Morocco, and loved it so much that she decided to stay.

Until she embarked on the Draa expedition, Alice's focus had been largely on getting from one point to another in the fastest possible time - not just on foot, but in the world's longest cycle race, the four-month Tour d'Afrique from Cape Town to Cairo. This new adventure demanded a different outlook and a slower pace. She would not be alone: her companions were Brahim Ahalfi, Brahim Boutkhoum and Addi Bin Youssef, respectively the expedition leader, guide and chef, and camel handler. All belonged to the indigenous Amazigh people who have long inhabited this part of North Africa.

After the initial greetings, Brahim Ahalfi gave Alice a new name in Arabic: Zahra, meaning 'flower.' He had brought five camels with him - all male, because mixing the sexes made them impossible to handle - and Alice tried to win their affections by feeding them tangerine peel, which they loved. Amid her excitement, her friend Jean-Pierre Datcharry, who had helped organise the expedition, gave her some sober words of advice: check your shoes for scorpions before putting them on; always wear sunglasses; look after your feet; zip up your tent against snakes, scorpions and ticks; keep clothes and bedding in plastic bags in case it rains; and be sure to send your GPS coordinates to the relevant authorities every evening.

Their first camp site was by the Mansour Eddahbi dam, whose reservoir of 560 million cubic metres is fed by rainfall and streams flowing from the Atlas Mountains, and provides water for the city of Ouarzazate. Before the dam was built in 1972, the Draa's upper and middle reaches flooded regularly, but now the flow is controlled in order to service Morocco's 'date basket' - the fertile oases that line its upper shores. Alice watched from her camp chair as the setting sun brushed the lake with gold, while the musical voice of Brahim Ahalfi intoned the evening prayer.

A friend who had been filming the expedition's departure went home after the first couple of days, leaving Alice to confront a secret anxiety. Physically, she knew that she was fit enough, but she was worried about how she might be perceived by her three companions. After all, they shared a different social and religious creed. How would they regard a single woman, far from her own family, travelling independently? Attacked by doubt, she lingered at the rear of the camel train. As if reading her thoughts, Brahim Ahalfi turned and shouted to her from the front. Why was she walking on her own, he wanted to know. One hand, he added, couldn't clap. Gratefully, she ran up to join him.

With the passing of time, life settled into a daily rhythm that was governed by natural light and the needs of the camels. The team got up at dawn, had breakfast, loaded the camels, and then walked for about five hours, with a mid-morning break for oranges and biscuits. Alice quickly learned that, once the animals got going, stopping was not an option: 'They maintain their pace whatever the terrain and if you stop for a pee or to take a photo it means you have to run to catch up.' [2] Every camp site was chosen carefully, well away from the tempting shade of palm groves where the camels would quickly devour any greenery.

In spite of Alice's initial fears, friendships began to grow. How could they not, with Brahim Boutkhoum baking a delicious cake iced with orange juice and vanilla sugar? The men jokingly gave Alice a new nickname, *Tafilfilat harran*, meaning 'hot chilli pepper,' because of her impatience. Addi, the youngest, couldn't resist a prank, pretending that the flesh of a bitter, spiny cactus tasted like the honey of wild bees and then dancing with uncontrolled mirth when Alice tried it herself.

Soon, talking became an integral part of the journey, so that with every footstep they were weaving words. Alice began to pick up phrases in Tashlaheet, the men's native tongue. Everything was discussed, from the English name for an Adam's apple to the difference in relationships before and after marriage. Addi regaled them with traditional stories that sometimes lasted for five days; in the evenings, after a dinner of soup eaten with bread and dates, followed by stew or kebabs, someone usually came up with a riddle. *(A city, surrounded by green gardens, with red houses, black inhabitants, and the key to it is a knife. The answer: a watermelon).*

Stopping for a few hours to visit friends in a village, the men took the opportunity to visit a mosque. In their absence, Alice was willingly led off by the women of the house and dressed up as a bride in jewel-coloured fabrics embellished with spangles and tassels; the bridal dress was then covered with a white gauzy wrap. Pleased with the result of their labours, her new friends then implored her to marry one of their menfolk and live with them forever. Alice began to wonder about her general philosophy of saying 'yes' first and asking questions later, but the men returned from the mosque in time for a feast of couscous and a good-humoured escape.

Other folk were encountered along the way: nomads, often with small herds of goats in tow, following a way of life that has remained largely unchanged for centuries. Warm hospitality lies at the heart of desert life, and they were eager to share their food and their time with fellow travellers. Ushered into their guest tents, Alice and her companions were offered sweet tea, fermented goat's milk, dates and freshly baked bread. Addi, himself a nomad, exchanged news of families and tribes. It was a time to relax and make new friends.

The baking heat lasted from dawn to dusk, but the nights were so cold that Alice developed open cuts between her fingers, which Brahim Ahalfi treated with raw lemons. A strong wind, arriving unexpectedly one morning, made her reach for her ski goggles to protect her eyes from blown sand. She carried with her a medical kit for treating snake venom, but knew that if any of them were bitten by certain species, there would be no help available.

All these discomforts and risks were outweighed by the simple joy of travelling. Adventurers, writes Alice, need three things: an optimistic nature, a short memory, and a big dollop of luck. Moving across the surface of the Earth, far from the noise and distractions of modern civilisation, she felt deeply at peace. The act of walking, she believes, has been hard-wired into human behaviour by our hunter-gatherer ancestors; the vast space allows us to breathe and to imagine. We remember the importance of being, as opposed to doing.

For the first section of their route, the terrain that Alice and her companions traversed was arid and stony; they had yet to reach the stereotypical desert landscape of rolling dunes. Alice was looking forward to it, and was surprised to find that the men were, too, despite the inherent difficulties of walking through deep sand.

When they reached Erg Chigaga, one of Morocco's largest ergs or sand seas, Alice and Brahim Ahalfi spent an hour scrambling to the top of a dune in order to gaze out at the sea of sand stretching away on every side. Here, the wind was so strong that they had to shout to make themselves heard. In a dusty valley that once contained water, drought-tolerant tamarisk bushes filtered the windblown sand so that little pyramids had built up against their roots.

The winding course of the Draa passes through one of the Earth's oldest inhabited areas, which is scattered with rock carvings and mysterious tumuli of earlier civilisations. Before the expedition, when Alice was writing to prospective sponsors, she optimistically added that she would be searching for lost cities, but she was astonished when she actually found one. On a rocky hilltop, her attention was caught by the remains of a two-chambered stone building which put her in mind of a crofter's house. Then she realised there were lots more. 'The ruins,' she wrote, 'stretched over the entire hill top, round the next peak and onto the next valley.' [3] Although not a true 'lost city' - after all, nomads had walked here for centuries - it was still an exciting discovery.

Traces of disused irrigation channels told a story of drought and abandonment.

The people who lived in these buildings, and the abundant wildlife depicted in early rock carvings, could only have survived in the presence of much more water than there is today. Traces of disused irrigation channels told a story of drought and abandonment. Alice and her companions walked for 25 kilometres across the desiccated bed of Lake Iriqui, 'so flat and so dry and so wide that you can see the curve of the world on the horizon.' [4] Until the late 1900s, the lake supported seasonal fishing and trading communities, and an abundance of wildlife. Alice was told that the last heavy rain fell in 2014, and afterwards the lake was a sea of grass and flowers.

For much of its length the Draa runs underground, surfacing here and there to allow an enterprising farmer to grow a good crop of beans. Often the only plants that Alice could find were tiny ones, tucked into crevices, surviving on dew. With her companions, she discussed the likely causes of this water scarcity: was it the damming of the Draa at Mansour Eddahbi? They agreed that it was possible, but the area's annual rainfall was also decreasing steadily, and this issue was affecting the whole of the Sahel. The discovery of human bones in an old graveyard, half-exposed by drifting sand, reinforced the impression of an increasingly lifeless environment. In the searing midday heat, they dug new trenches and re-buried them respectfully.

Finding water for themselves was becoming a much more serious consideration, especially when some of the wells they visited turned out to be dry. 'There was something devastating,' wrote Alice, 'about opening the well cover to look down and see nothing but sand.' [4] Longer and longer detours were necessary, just to find sufficient water; and then, after plunging a bucket rather too vigorously into a well, Alice lost the bucket, and even Addi objected to the idea of being lowered down to retrieve it. Luckily they had other containers, but the nightmare of a waterless situation suddenly seemed very close. As they continued their journey, the men teased Alice about the mishap at regular intervals, but all of them were astonished when they came across an old black bucket - thereafter known as 'the miracle bucket' - very similar to the one they had lost, lying abandoned in the desert.

'There was something devastating about opening the well cover to look down and see nothing but sand.'

In the 40-degree heat, Alice began to feel the stress that comes from physical exertion and tiredness. Staying hydrated was now much more of a challenge. Her patience began to fray, but the men were so kind to her that she could find nothing to grumble about, which made her even more irritable. Addi, at least, had a focus for his own frustration. His mobile phone, an ancient model that he'd bought cheaply in a market, had been working intermittently but now gave up the ghost altogether. Coming back from tending the camels, Brahim

Ahalfi dropped something into his lap: a stone, curiously shaped like a mobile. Restored to good humour, Addi 'charged' it with a twig.

Having travelled southwards for most of the trek, their steps turned westwards towards the Atlantic. One morning, Alice opened her tent to find mist blowing inland from the sea. She enjoyed its freshness on her skin, and discovered that other forms of life thrived in it too, not all of them welcome. Her tent was soon crawling with insects that the men called *Ami Wislan*, a name meaning 'my paternal uncle with two claws'.

And then there were ticks, fat ones, so numerous that they made the ground shimmer. Alice hastily tucked her trousers into her socks, but the camels were defenceless and miserable, with ticks in the corners of their eyes and behind their ears. With Brahim Ahalfi holding them fast in their halters, they roared in protest as Addi painted their skin with a protective oily tar made from the bark and sap of juniper.

As they approached Oued Chbika, near where the much-reduced Draa flows into the sea, Alice was dreading her journey's end. Over a total of 81 days, she had grown addicted to the natural rhythm of desert travelling, where life is whittled down to the basics and the world seems fresh and new. 'In the great calm of the wilderness,' she wrote, 'you have time to see and hear everything: a tiny wildflower in a tumble of rocks, the halting song of the black and white mulla mulla bird, the flash of a fat green caterpillar undulating along the sand.' [5] The journey was, she believed, one of the best things she had done in a life that was already full of wonder. Saying goodbye to her companions was going to be hard.

But just before the end came the prospect of a new beginning. Over tea and another of Brahim Boutkhoum's fragrant cakes, Alice met expedition organiser Jean-Pierre Datcharry to discuss two further journeys, one taking her south from Oued Chbika and another leading from the Mediterranean coast to Ouarzazate. It was possible that some, if not all, of her team from the Draa adventure could go with her. If they completed these new routes, they would have walked across the whole of Morocco and the Sahara.

It was an enticing prospect, one that fed Alice's impulse to keep moving and exploring. Meanwhile, she still had some final steps to take. At Oued Chbika, the four desert travellers crested the last low hill and led the camels onto the beach. Then they linked arms and sang with joy as they walked into the sea.

Alice subsequently completed the second and third stages of her journey, which came to an end in October 2020. Walking, she says, is very slow, but if you keep going, you will get there - a wonderful metaphor for life, and one which she continues to apply to her own adventures.

[1] Alice Morrison, 'The Draa Expedition - The Lost City' https://www.youtube.com/watch?v=ft1WIBlyobk

[2] Alice Morrison, 'Draa Expedition: First Steps' (30th January 2019) https://alicemorrison.co.uk/draa-expedition-first-steps/

[3] Alice Morrison, 'Indiana Alice and the Lost City' (12th February 2019) https://alicemorrison.co.uk/indiana-alice-and-the-lost-city/

[4] Alice Morrison, 'Where has all the water gone?' (24th March 2019) https://alicemorrison.co.uk/where-has-all-the-water-gone/

[5] Alice Morrison, 'Journey's End' (16th May 2019) https://alicemorrison.co.uk/journeys-end/

OCEAN

Studying the depths of the sea is like hovering in a balloon high above an unknown land which is hidden by clouds.

Sir John Murray, *The Depths of the Ocean*

But where, after all, would be the poetry of the sea were there no wild waves?

Joshua Slocum, *Sailing Alone Around the World*

E ola ke kai, E ola kākou
(As the ocean thrives, so do we)
Hawaiian proverb

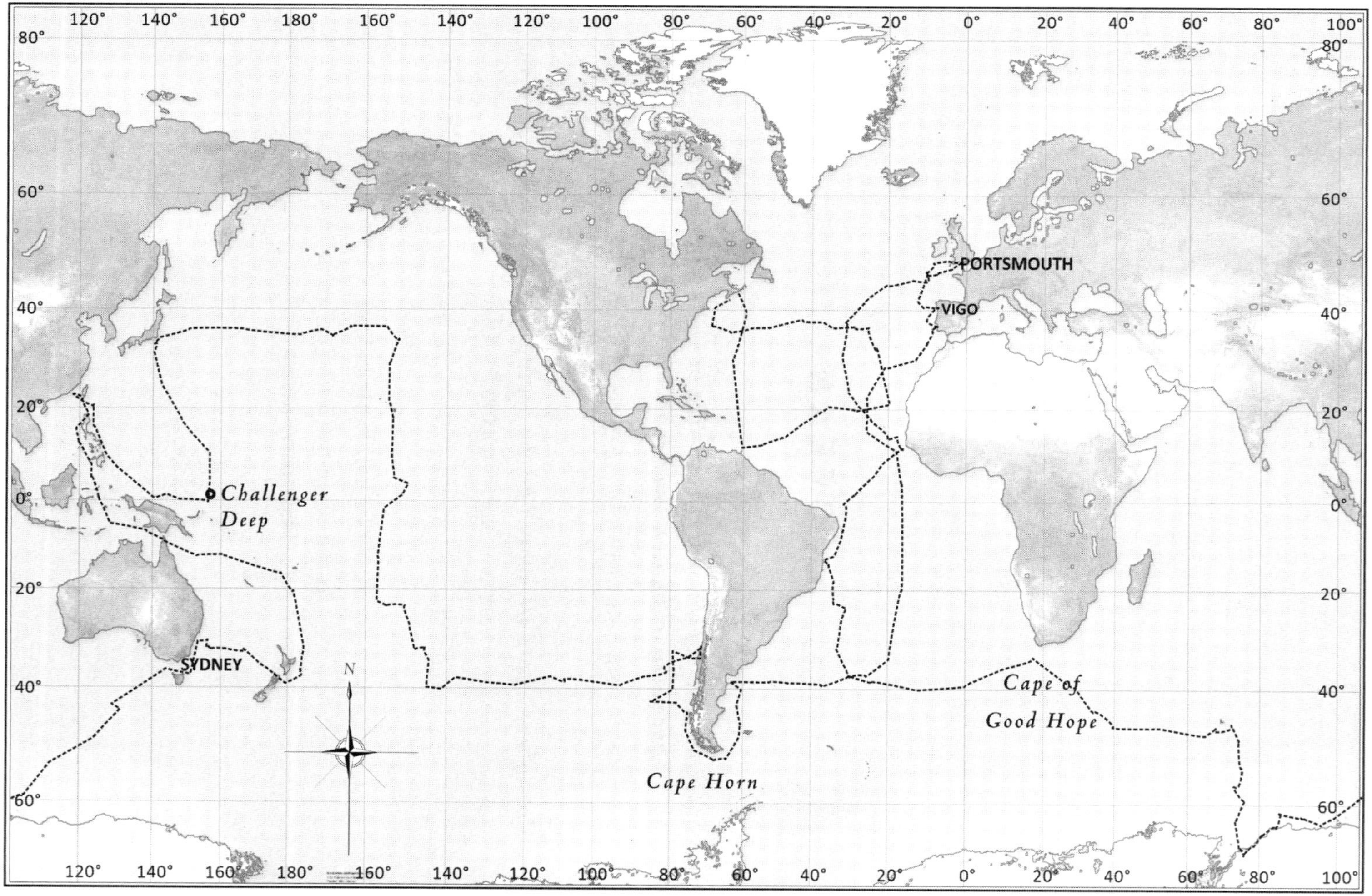
PORTSMOUTH
VIGO
Challenger Deep
SYDNEY
Cape of Good Hope
Cape Horn
N

Sir Charles Wyville Thomson

o

'The mud!' wrote Sub-Lieutenant George Campbell. 'Ye gods. Imagine a cart full of whitish mud, filled with the minutest shells, poured all wet and sticky and slimy onto some clean planks... In this the naturalists paddle and wade about, putting spadefuls into successively finer and finer sieves till nothing remains but the minute shells &c.' [1]

The 'clean planks' that were being soiled with mud belonged to HMS *Challenger*, a warship of the Royal Navy that had been re-purposed as a floating oceanographical laboratory. At the time of Campbell's writing, which was January 1873, she was just one month into a four-year expedition around the world. Deep-sea dredging, it turned out, was a slow and messy business, but the scientists on board cared only for the wealth of marine life that was being hauled out of the abyss and deposited onto a specially-built platform in bulging, leaking nets. For those officers who liked to keep a clean ship and enjoyed a fast-paced life on the high seas, it was going to be a very long voyage.

In charge of this oceanic mud-bath was a Linlithgow-born scientist named Charles Wyville Thomson. As a young man, like many students of his day, he had progressed from medicine to an interest in natural science, and by the time he was appointed Chair of Natural History at Edinburgh University he had already held several eminent academic posts in Scotland and Ireland. In the late 1860s, in the company of zoologist William Carpenter, he had used a couple of Royal Navy vessels to undertake scientific explorations in the North Atlantic.

Thomson's initial voyages were necessarily short, but he and Carpenter found evidence to suggest that marine creatures were living at depths of of 2,000 fathoms (3,658 m). This discovery contradicted the theory of Edward Forbes, one of Thomson's predecessors in his Edinburgh professorship. Forbes had proposed that the abundance of life decreased in inverse proportion to the ocean's depth. Discussing what was then thought to be the deepest life-supporting region, that of deep-sea corals, he wrote: 'As we descend deeper and deeper in this region its inhabitants become more and more modified, and fewer and fewer, indicating our approach towards an abyss where life is either extinguished, or exhibits but a few sparks to mark its lingering presence.' [2] Forbes believed that no life at all could exist below 300 fathoms, and this apparently inhospitable region was named the azoic zone.

In his book, *The Depths of the Sea*, Thomson politely refuted Forbes' theory with accounts of his own first-hand experiences: 'The bed of the deep sea,' he wrote, '...is not a barren waste. It is inhabited by a fauna more rich and varied on account of the enormous extent of the area, and with the organisms in many cases apparently even more elaborately and delicately formed, and more exquisitely beautiful in their soft shades of colouring and in the rainbow-tints of their wonderful phosphorescence, than the fauna of the well-known belt of shallow water teeming with innumerable invertebrate forms which fringes the land.' [3]

'The bed of the deep sea is not a barren waste.'

And there was another scientist for whom life in the deep sea held out a significant promise. In 1859, with the publication of his book, *On the Origin of Species,* Charles Darwin advanced his ground-breaking but controversial theory about evolution by natural selection. Darwin believed that a host of marine creatures preserved only as fossils on land might still be flourishing, as yet undiscovered, in the depths of the ocean. If the evolution of species could be portrayed graphically as a tree, maybe the sea would yield some of its missing branches.

So were the deep oceans entirely devoid of life, or did they hold the key to the existence of all complex organisms on Earth? Thomson disbelieved the former hypothesis, and he was eager to investigate the latter. He explained: 'I had long previously had a profound conviction that the land of promise for the naturalist, the only remaining region where there were endless novelties of extraordinary interest ready to the hand which had the means of gathering them, was the bottom of the deep sea.' [3]

While passenger-carrying bathyscaphes were still a thing of the future, Thomson had already proved the effectiveness of sounding lines and dredges that could be lowered from a ship. On a dedicated scientific voyage, using a specially-equipped ship as a floating laboratory, he imagined plumbing the world's oceans and gathering all kinds of data for analysis. With his reputation and his network of contacts, he was ideally placed to set it in motion.

As a Vice-President of The Royal Society, Thomson's former seafaring companion William Carpenter was a powerful ally. Together, they persuaded the Admiralty to lend them a vessel, and the British government agreed to finance an expedition. There was an extra incentive, in fact: with the advent of telegraph communication, it would benefit Royal Navy's hydrographers to know the depth of an ocean before laying cables across its bed.

Funding was approved in April 1872, and by December of that year a Royal Navy warship, stripped of all but two of her guns and converted into a fully-functioning research vessel, was ready to depart. HMS *Challenger* was a hybrid of sail and steam, a 200-foot, three-masted corvette fitted with an auxiliary steam engine. She carried a company of 269 men, made up of officers, crew and scientists. For all, it was a huge commitment: they were unlikely to see friends and family again for the best part of four years.

Commanding the ship was Captain George Nares, an experienced Royal Navy officer who had taken part in the search for Sir John Franklin's lost expedition in the Arctic. Thomson's team of scientists included John Murray, a Canadian-born man of Scottish descent who had interrupted his medical training at the University of Edinburgh with a spell as ship's surgeon on an Arctic whaling cruise. Murray had returned to Edinburgh to study geology under Sir Archibald Geikie, and he assisted Thomson in preparing scientific apparatus for *Challenger* before being confirmed as a member of the expedition team.

The others were Henry Nottidge Moseley, who had a first-class degree from Oxford's Honour School of Natural Science; a German naturalist called Rudolf von Willemoes Suhm, who was an assistant in the zoological museum at the University of Munich; John Young Buchanan, a Scot who had studied chemistry at Glasgow University and was assistant to Alexander Crum Brown at the University of Edinburgh; and John James Wild, a Swiss-born linguist and illustrator who was appointed as Thomson's secretary and the expedition's official artist.

Challenger left Portsmouth harbour on 21st December and sailed straight into a gale in the Bay of Biscay. She rolled heavily for days, making it impossible for anyone to stand up, let alone enjoy their Christmas dinner. Crockery was smashed, sleep was near-impossible, and their morning coffee, reported Sub-Lieutenant George Campbell, went down their boots instead of their throats. Thomson merely observed that the rough

Their morning coffee went down their boots instead of their throats.

sea greatly hindered their first attempts to lower a dredge; they were still having difficulty when they reached Lisbon, where some of the seasick officers were glad to go ashore and do some sightseeing. They would, Thomson explained, 'consider every thing done during the voyage from England to the Canary Islands as tentative and introductory.' [4]

Sounding and dredging work began in earnest when they set out across the Atlantic. 'Hundreds of miles of line,' wrote Thomson, 'are reeled and coiled in every available spot on the forepart of the main-deck and elsewhere.' [4] Depth-sounding involved letting down a weighted instrument called a Baillie sounding device and measuring, by 25-fathom marks on the line, how much was paid out before it reached the bottom. The depth-sounder brought back a small sample of water and sediment, and it was usually accompanied by a thermometer; during this procedure, the ship's engines were used to maintain a constant position.

Then a dredge was let down. This device comprised a large net, made of two different grades of mesh, attached to a frame that was designed to hold the net open as the ship dragged it slowly across the sea bed. A small engine installed on deck assisted in hauling up the heavy load. In the deepest water, the work of lowering, filling and raising the dredge occupied all the daylight hours, and the scientists studied their finds under microscopes as the ship sailed towards her next sounding point. Samples of water were taken from varying depths, to provide as complete a picture as possible of conditions between the surface and the sea bed.

Commonly, what the dredge deposited on deck was a type of soft mud known as ooze, which accumulates on the floor of the deep sea and contains the skeletal remains of dead micro-organisms. Oozes are classified according to their constituents; *Challenger*'s dredges were repeatedly yielding two kinds of calcareous ooze, which the scientists named *Pteropod* ooze (consisting of tiny molluscs) and *Globigerina* ooze (containing a genus of marine plankton).

While this ooze-cruise was a source of delight and fascination for the scientists - or 'philosophers', as some of the officers called them, because their attire was more suited to a stroll along Pall Mall - it soon became evident that the rest of the ship's company found the voyage insufferably boring. Captain Nares persuaded Thomson to give talks to the crew about their findings, which inspired a certain amount of interest; but the leisurely rate of progress was still too much for some to endure. During the voyage as a whole, about a quarter of the crew deserted when they reached a convenient port. Some were tempted to try their fortunes gold-prospecting in Australia, while others headed for the diamond mines of South Africa.

As *Challenger* crossed and re-crossed the Atlantic, Thomson and his team made one of their most significant discoveries: the chain of 'mountains' forming the mid-Atlantic Ridge, which is now known to be the boundary of several tectonic plates. Eventually the ship worked her way down to Simonstown and the Cape of Good Hope, where the sun's heat was so intense that the crew had to throw water on her decks to stop the pitch from melting.

But the summer sun became a distant memory when they headed south towards the Antarctic. Heavy waves in the Roaring Forties tossed the ship around like a cork, and the wind chilled her occupants to the bone. On Christmas Day 1873, the second of four Christmases spent on board, the crew assembled on deck and sang 'The Roast Beef of Old England' before descending, perhaps with pangs of nostalgia, to a simple dinner of salt pork and pea soup. The officers probably fared slightly better, but 'Oh,' wrote Sub-Lieutenant Campbell with deliberate irony, 'the pleasure of keeping watch on wet decks, with sharp hail and sleet pelting one's face and freezing one's toes.'[1]

'The pleasure of keeping watch on wet decks, with sharp hail and sleet pelting one's face and freezing one's toes.'

On 16th February 1874, on the edge of the Great Ice Barrier (now called the Ross Ice Shelf), *Challenger* reached her 'furthest south', a latitude of 66°44'S. She was only 1,400 miles (2,253 km) from the South Pole. Not long afterwards, with a dredge freshly hauled up in the teeth of a southerly wind, she strayed too close to an iceberg and collided with it. The jib-boom was carried away, and the riggers fought to save the topgallant mast while stokers fired up the engines to get the ship clear.

But still they weren't out of danger. In the midst of a squall, a shout came from the forecastle: "Iceberg close to under the lee bow, Sir!" followed by a cry of "Full speed astern!"[1] As another giant berg loomed up across their path, Captain Nares shouted orders from the bridge, while the men worked desperately to shake loose the main topsail. Slowly, *Challenger* by-passed the berg, but after that there was no peace to be had, and the officers were exhausted by the need to keep a constant vigil. It was with relief that they reached ice-free waters once more, and headed north towards Australia, New Zealand and the Pacific.

Thomson was still keenly searching for 'missing links' in Darwin's evolutionary tree. After a foray into the interior of Queensland, Australia, he returned to the ship laden with botanical specimens and a peculiar looking fish which excited a great deal of attention. Lungfish, which are also present in South America and Australia, use lungs to supplement their gills and have rudimentary limbs for walking on land; they are the only surviving members of the *Sarcopterygii* or lobe-fins, believed to represent a key development in the evolution of tetrapods which began about 400 million years ago. Scientists still debate the precise evolutionary details, but Thomson and his scientists recognised the significance of this extraordinary creature.

On 23rd March 1875, nearly two weeks out of Papua New Guinea, the sounding line reached an astonishing 4,475 fathoms (8,184 m). Two thermometers, sent down at the same time, were found to be smashed when they came back up. *Challenger* was directly above the deepest known point on the Earth's seabed, a staggering abyss that is now known to lie 5,969 fathoms (10,916 m) below the sea's surface at its deepest point. In honour of the expedition, it was named the Challenger Deep.

Early in May 1876, a few days north of the Azores, *Challenger* paused at her 354th (and last) deep-sea observing station before setting sail for home. Persistent north-easterly winds forced her into the Spanish port of Vigo, where the entire Channel Fleet of Royal Navy ships was assembled. 'We had all many friends in the fleet,' recalled Thomson, 'and much to say and hear.' But, he added, on *Challenger's* arrival, the band on HMS *Defence* 'struck up the air "Home, Sweet Home", and tried the nerves of some of us.' [4] A change in the wind hastened their departure; almost before they knew it, they were across the Bay of Biscay and powering up the Solent. Thomson's record-taking was consistent to the last. He wrote: 'After an absence of three years and a half, we stopped and came to an anchor in seven fathoms' water at Spithead.' [4]

Sadly, seven lives had been lost since their departure, including Rudolf von Willemoes-Suhm, who died of a bacterial infection and was buried at sea. At least a further 20 men had been invalided off the ship because of sickness.

During their voyage of more than 68,000 miles (109,000 km), Thomson and his scientists collected or observed some 4,700 species that were new to science, including about 10% of all known starfish. A species of brachiopod, found at a depth of more than 15,700 feet

(4,785 m) and named *Abyssothyris wyvillei* after Charles Wyville Thomson, was just one of a myriad creatures whose existence proved conclusively that life existed in the deepest parts of the ocean.

But had Thomson managed to find Darwin's 'living fossils', those eagerly hoped-for deep-sea organisms that were preserved only in rocks on land? Despite his impressive list of newly-discovered species, it was generally considered that he had not. James Geikie, Professor of Geology at the Edinburgh University, was quite clear: 'The results of the expedition afford no justification for the assumption that a universal and peculiar fauna of great antiquity occupies the deep sea…'[5]

Thomson was of the same mind, but it seems that a lot hinged on individual interpretation. In his own account of the voyage, Thomson wrote: 'The character of the abyssal fauna refuses to give the least support to the theory which refers the evolution of species to extreme variation guided only by natural selection.'[4]

This was too much for Charles Darwin, who retorted: 'I am sorry to find that Sir Wyville Thomson does not understand the principle of natural selection, as explained by Mr [Alfred Russel] Wallace and myself… Can Sir Wyville Thomson name any one who has said that the evolution of species depends only on natural selection?'[6]

Thomson might have been forgiven for taking some much-needed rest in the company of his wife, Jane and his son, Frank Wyville, a medical student. Instead, newly-knighted and with a clutch of academic awards to his credit, he threw himself into the task of setting up the Challenger Expedition Offices in Edinburgh, where work began on cataloguing the collection of over 100,000 specimens. Years of over-exertion, however, had taken a severe toll on his health, and in March 1882 Thomson succumbed to the effects of paralysis, generally attributed to exhaustion. He was 52.

Tribute was paid to Thomson's ability as a lecturer, which raised the status of zoology as an academic subject, and to his popularity among University colleagues and students. Meanwhile, one of his fellow scientists on the *Challenger*, John Murray, inherited the monumental task of cataloguing the collections and publishing the official scientific report, which was only completed in 1895 and ran to 50 volumes. Many of the specimens are now housed at the Natural History Museum in London, while the Royal Museums in Greenwich hold archival material and images.

Thomson's legacy persists to the present day. The Challenger Expedition marked the birth of modern oceanography, and has been called the Apollo mission of the Victorian era. Data and drawings brought home by *Challenger*'s scientists preserve a record of the oceans from over 150 years ago, and provide us with a benchmark for comparing modern-day temperatures, currents, salinity and composition, as well as flora and fauna. Photographs of Antarctic icebergs, captured in 1874, are believed to be the first ever taken. Among the more curious deep-sea finds were micro-meteorites, tiny spherules of cosmic dust; these particles are raining down on Earth all the time, but this was the first time they were detected.

It would no doubt have astounded and delighted Thomson to know that, in 1960, Swiss oceanographer Jacques Piccard and US Naval Lieutenant Don Walsh returned safely from a dive of nearly seven miles (11.3 km) to the bottom of the Challenger Deep. Their descent to the ocean floor took nearly five hours. The window of their bathyscaphe cracked under the pressure, but fortunately, unlike Thomson's thermometers, it did not implode.

[1] Lord George Campbell, *Log-Letters from the Challenger* (1881)

[2] Edward Forbes, *The Natural History of the European Seas* (1859)

[3] C Wyville Thomson, *The Depths of the Sea* (1873)

[4] C Wyville Thomson, *The Voyage of the Challenger* (1877)

[5] James Geikie, 'The Challenger Expedition', *Scottish Geographical Magazine* (May 1895)

[6] Charles Darwin, 'Sir Wyville Thomson and Natural Selection', *Nature* (11th November 1880)

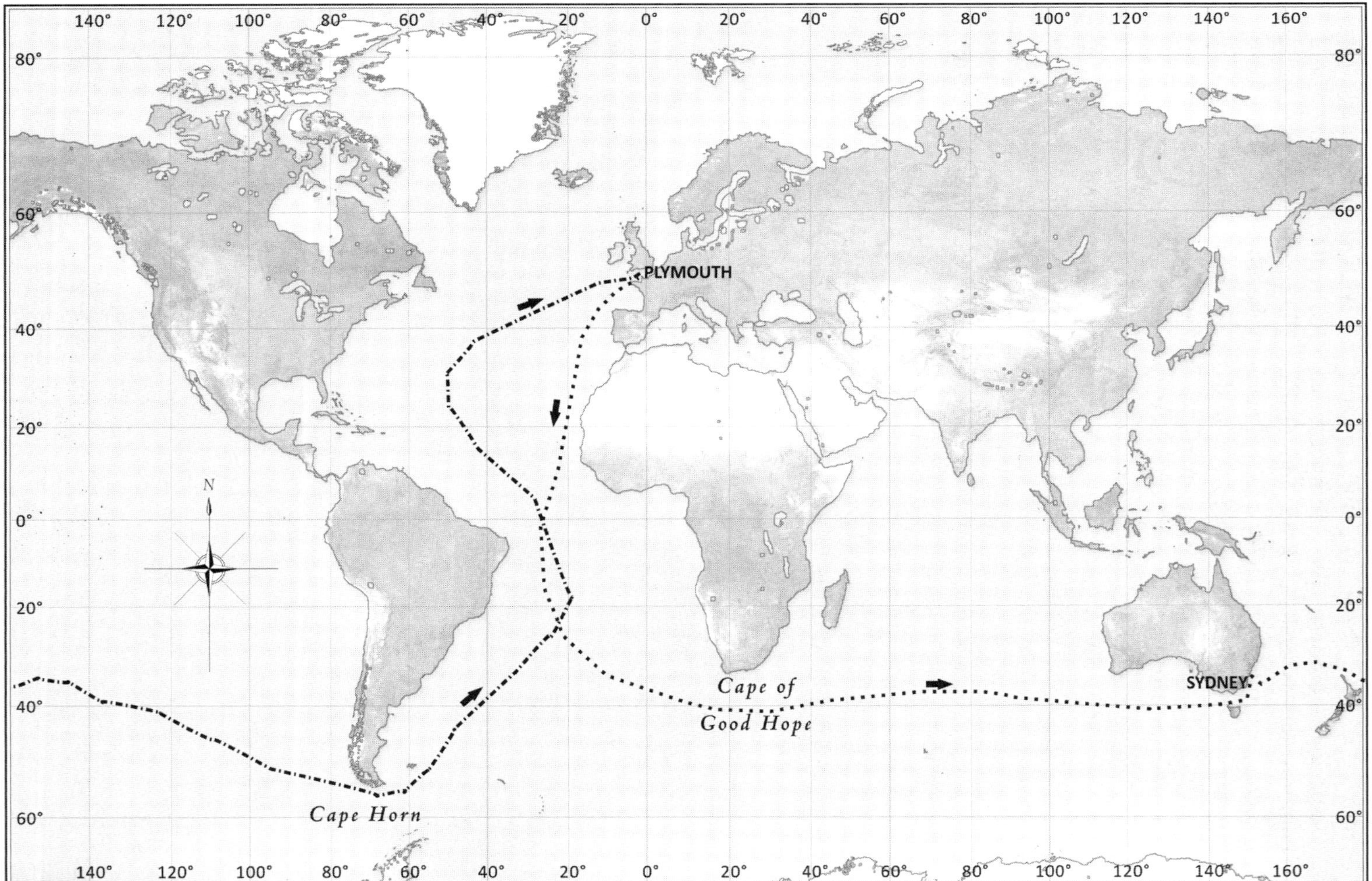
PLYMOUTH
SYDNEY
Cape of
Good Hope
Cape Horn
N

Sir Francis Chichester

In September 1953, around the time of his 52nd birthday, a man from North Devon telephoned his wife and told her rather nervously that he'd just bought a boat. He hadn't done a great deal of sailing, but he wanted to enter some ocean races to see if he could learn something. His name was Francis Chichester.

Deep down, Francis was cherishing a long-held dream of circumnavigating the world. As a young man, he'd thought of doing it by air, and embarked on some rather risky solo adventures in a De Havilland *Gipsy Moth*. In the absence of a compass, he used railway lines to navigate - a method known as 'flying by Bradshaw' after the well-known railway guidebooks. On one occasion, he checked his accuracy by circling low over a train station and reading the name on the platform. To his delight, he was spot on.

Francis was serious about honing his aviation skills, practising emergency landings by cutting the engine at 1,000 feet (305 m) and then picking a suitable field. He learned how to make essential repairs, measured his petrol consumption at different speeds, and fitted a compass which allowed him to navigate at night. He felt that he was ready for a big adventure; by way of a trial run, he made an ambitious loop around Europe, getting as far east as Iași in Romania and Chernivtsi in Ukraine. Then, in December 1929, he took off from Croydon aerodrome and headed for Australia.

It wasn't as if he was oblivious to fear. But he chose to focus his restless energy on solving logistical problems. Airfields had to be located, permits obtained for over-flying certain countries, fresh petrol supplies made available along his route. He collected all the relevant maps and cut them into long strips, so that he could use them while flying. He made it to Sydney in 41 days, having been forced to wait in Tripoli for a new propeller, and surviving a day and a night in the Australian outback with only three gallons of petrol and no exact idea where he was.

> *'No one had flown across the Tasman Sea alone, and I had a great urge to be the first to do it.'*

In Sydney, he was greeted like a hero, but he found it hard to cope with the attention. A new idea had entered his head: 'No one,' he wrote, 'had flown across the Tasman Sea alone, and I had a great urge to be the first to do it.' [1] There was one major problem: the 1,200-mile (2,000-km) distance was too great for a non-stop flight in his little plane. He would have to do it in a wide arc, leap-frogging between islands. Because there were no suitable landing sites, he fitted floats to the *Gipsy Moth* so that he could alight on the sea.

Almost crashing off Lord Howe Island in a gale, the plane collapsed into a shallow lagoon and sank in shallow water. Helped by willing islanders, Francis set about salvaging and re-building her, using spare parts sent from Sydney. Navigating by sextant, he made it across the Tasman Sea, and turned his attention to completing the second half of his round-the-world flight. It was then that his luck changed. Taking off from the harbour at Katsuura in Japan, he flew into some telegraph wires and his dreams broke into a thousand pieces.

Twenty-two years later, he was the proud new owner of a yacht called *Florence Edith*. He re-named her *Gipsy Moth II*, after his beloved little plane.

Francis had been toying with the idea of yachting for a while. During the war, he had taught navigation techniques to RAF pilots, and more recently he had advertised his services as a navigator to boat owners looking for crew. But, he wrote, 'nobody seemed interested in an air navigator who knew nothing of the sea. So I was forced to buy a yacht of my own in order to learn.' [1] And he was serious about learning. He joined the Royal Ocean Racing Club, found some crew, and entered the North Sea Race, a course of 220 miles (354 km) from Harwich to Rotterdam. He ran aground at the start and arrived nearly last; undeterred, he simply entered more races.

To begin with, some of Francis's crew expressed private doubts. While admiring the originality of their skipper's mind, they were less sure about the strain he was putting on his body, and they were alarmed by the way he used air navigation methods to pilot a boat. How, they wondered, would checking the night sky keep them off the sandbanks? Once, when Francis had gone below for a sleep, one of them remarked, 'Do you think he knows where we are?' But after four seasons, Francis knew enough to realise that there was a lot more to be learned about ocean sailing. It was time to get a bigger boat; he turned his attention to designing a new one, which he called *Gipsy Moth III*.

The first ever single-handed trans-Atlantic race was the brainchild of Herbert 'Blondie' Hasler, himself a yachtsman and an eager contestant. After several years of planning, the first race was scheduled for 1960. For Francis, it could not have come at a better time. He had spent a few years battling ill health, and the prospect of crossing the Atlantic solo came as a much-needed boost to his recovery. Although he'd never sailed alone in a vessel bigger than a 12-foot dinghy, his imagination was fired and he set about designing a self-steering device so that he would be able to sleep during the voyage.

Five boats entered the race, and the starting gun was fired in Plymouth on 11th June. After three days, Francis was only 186 miles to the south-west. It was rough weather: his muscles ached from hoisting and lowering the large sails, he was bruised from being thrown from one side to the other, and he was soaked to the skin by waves sluicing across the deck. Then the boat started hitting her stride, cutting through the water so fast that it was hard to stand up. Francis wrote that he would not change places with anybody in the world.

On 21st July, a distinctive white sail was glimpsed on the horizon from Long Island, New York. Within hours, *Gipsy Moth* was surrounded by launches filled with photographers. Francis shouted to ask about the others. 'You're first!' they replied.

When a sponsor offered to help Francis build an even better boat for ocean racing, he knew exactly what he would do. The dream of his youth was still with him. He wanted to complete the fastest solo voyage around the world, following the route of the iconic tea clippers around Cape Horn. Strong winds and tremendous ocean currents make the seas around Cape Horn, off the southern tip of South America, among the most dangerous in the world. In 1966, Francis knew of only eight yachts that had rounded the Horn on a circumnavigation, and of these, six had capsized or somersaulted. The prospect terrified him, but he hated being held back by fear. This, he believed, was one of the greatest challenges left in the world. It remained to be seen whether he was up to it.

This, he believed, was one of the greatest challenges left in the world.

Gipsy Moth IV was the first ever purpose-built ocean racer. At 53 feet, with 854 square feet of sail (extending to 1,500 square feet with the spinnaker), she had been designed for speed, but she would be a handful for anyone to sail single-handed. In sea trials, Francis had been excited when she achieved nine knots easily in moderate winds, but alarmed at how much she heeled over. Last-minute adjustments and increased construction costs stressed him severely, and then he slipped on a skylight and injured his leg.

One thing was clear: there might be setbacks, but nothing was going to stop him. Meanwhile, something about this unconventional sailor was beginning to catch the public's imagination. At a pre-departure press conference, Francis was touched by the goodwill expressed by people up and down the country. On 27th August, the job of tacking to and fro before the starting line kept his mind sharply focused while he waited for the 11 o'clock gun, and then he was off down Plymouth Sound with 14,000 miles (22,530 km) of ocean between him and Sydney, his only planned port of call.

In the days before GPS, Francis was relying on charts, compass and sextant for navigation. He also realised very soon that the habits of the old clipper captains were worth studying. Within eight miles of Madeira, *Gipsy Moth* was struck by squalls funnelling down from the island's peaks, and suddenly she was out of control, tearing through the water while Francis hauled desperately on an unmovable tiller. Only when he'd got the mainsail and mizzen down, and escaped the island's wind shadow, did he have time to recall that the clippers always steered well clear of Madeira.

For his 65th birthday, Francis allowed himself a celebration. He donned his green velvet smoking jacket, made himself a champagne cocktail and indulged in fond thoughts of home. His wife, Sheila, was a staunch supporter of his solitary wanderings across oceans. She knew that he found healing in the peace of the sea. Not everyone was so perceptive. 'People keep at me about my age,' he wrote. 'I suppose they think I can beat age. I am not that foolish. Nobody, I am sure, can be more aware than I am that my time is limited... Our only purpose in life, if we are able to say such a thing, is to put up the best performance we can - in anything, and only in doing so lies satisfaction in living.' [2]

Five days later, *Gipsy Moth* crossed the Equator. By now, it was too hot to stay on deck for long in the daytime; to keep cool, Francis doused himself with buckets of seawater in the cockpit. Within a couple of weeks he was battling squalls of hail in the Roaring Forties, and it was then that the self-steering chose to break. For three days, with seas repeatedly slamming into the yacht, he struggled to make an effective repair.

On 12th December, *Gipsy Moth* sailed into Sydney Harbour. After three months of solitude, within ten minutes of stepping ashore Francis was ushered into a press conference crammed with reporters. He desperately needed sleep, but found himself bombarded with invitations that he had neither the will nor the energy to accept. Most of all, he was keen to get repairs and adjustments made to his boat, so that he could be on his way. To some observers he appeared thin and strained, and newspapers began to publish calls for him to end his journey now. He ignored them. On a fine Sunday morning, carrying a gift of three miniature bales of wool in honour of the clipper ships and their cargoes, he sailed out of the harbour amid a flotilla of little boats. Two nights later, *Gipsy Moth* capsized in the Tasman Sea.

According to radio reports, a tropical cyclone was passing just under 300 miles to the east. As winds began to rise, Francis reefed in the sails to a bare minimum and went to lie in his bunk, hoping to sleep. In pitch darkness, the sensation of rolling woke him up. He was more curious than frightened: the boat had been designed to be self-righting, and he wondered if she would roll over completely. At the same moment, crockery, cutlery and bottles began to rain down on his body with a deafening clatter. He had the oppressive feeling that the boat was on top of him. Then, quietly, she righted herself on the same side she had gone down. Francis reached for his bunk light, and was relieved to find that it worked. What it revealed was a scene of utter chaos.

From a mark made on the cabin roof by a whisky bottle, Francis later calculated that the boat had rolled to an angle of 131 degrees. A damage inspection revealed that one of the stored sails had gone but the mast and rigging were intact. Below deck, lockers had emptied themselves of their contents. The cabin, awash with water, contained a deep rubble of tins, bottles and tools. A knife had pierced his bunk, narrowly missing his head. As the storm eased, he started pumping out the bilges and contacted Sydney by radio to say that he had capsized, but that he was all right and didn't need help. A massive clean-up job lay ahead of him, and he still had to get around Cape Horn.

As strong winds propelled him across the Pacific, Francis tried to stow everything safely away in case the boat rolled again. He wondered what he would meet in the Southern Ocean: sailing friends had warned him about giant waves in the Drake Passage, where the Pacific and Atlantic Oceans converge, and there was an additional risk of icebergs. Over an intermittent radio connection, he heard from the British Antarctic Survey that there was no ice above 67 degrees south. Plagued by near-perpetual cloudy skies, he fought to get sun readings; here, of all places, he couldn't afford to make navigational errors.

Approaching Cape Horn, the fickle wind alternately veered and backed. Setting the self-steering and trying to sleep in such conditions was a test of nerve, but it was next morning, after he'd sighted the towering black cliffs of the Cape, that the ocean became truly ferocious. Waves towered 50 feet above him like grey-green walls of earth; *Gipsy Moth* rode them and dipped into their troughs. The cockpit filled with water five times, and the wind-reading machine stopped working. There was no option but to let the boat run before the wind, and trust that its direction would stay constant. Eating was almost impossible; he'd had only two meals in two days, supplemented by liquid snacks. He admitted that he was frightened.

A couple of days after passing the Horn, *Gipsy Moth* headed out into the the South Atlantic, followed by a retinue of albatrosses to which Francis fed some stale cake. It was raining and cold, and the wind was still strong, but the nightmare was over. Now he could set a course for home.

When *Gipsy Moth* eased her way into Plymouth Sound on 28th May, hundreds of thousands of people had turned out to greet her. The crew of the Royal Navy aircraft carrier HMS *Eagle* lined up on deck to give three cheers, a gesture which touched Francis deeply. To his astonishment, a replica *Gipsy Moth* biplane, exactly like the one he had piloted to Australia, flew overhead and dipped its wings in salute. Kneeling before Queen Elizabeth II a few weeks later, Francis was knighted with the same sword that was used to honour Sir Francis Drake when he returned from his own circumnavigation nearly 400 years earlier.

With his voyage, Francis had either broken or established many records: he was proudest of having made the fastest time around the world by any small vessel: 29,630 miles (47,685 km) in a sailing time of 226 days. He had also narrowly beaten the average clipper time of 230 days to Sydney and back. On a deeper level, he had rebelled against the apparent inevitabilities of life, and found exhilaration. The accolades were incidental; for him, the action was what mattered. He wrote: 'The Buddha said that life was a gift which any man can take but none can give; I think it is wrong not to use such a gift to the utmost, wrong to bury it like an unused talent, wrong not to use it to the very end.' [3]

[1] Francis Chichester, *The Lonely Sea and the Sky* (1964)

[2] Francis Chichester, *Gipsy Moth Circles the World* (1967)

[3] Francis Chichester, *The Romantic Challenge* (1971)

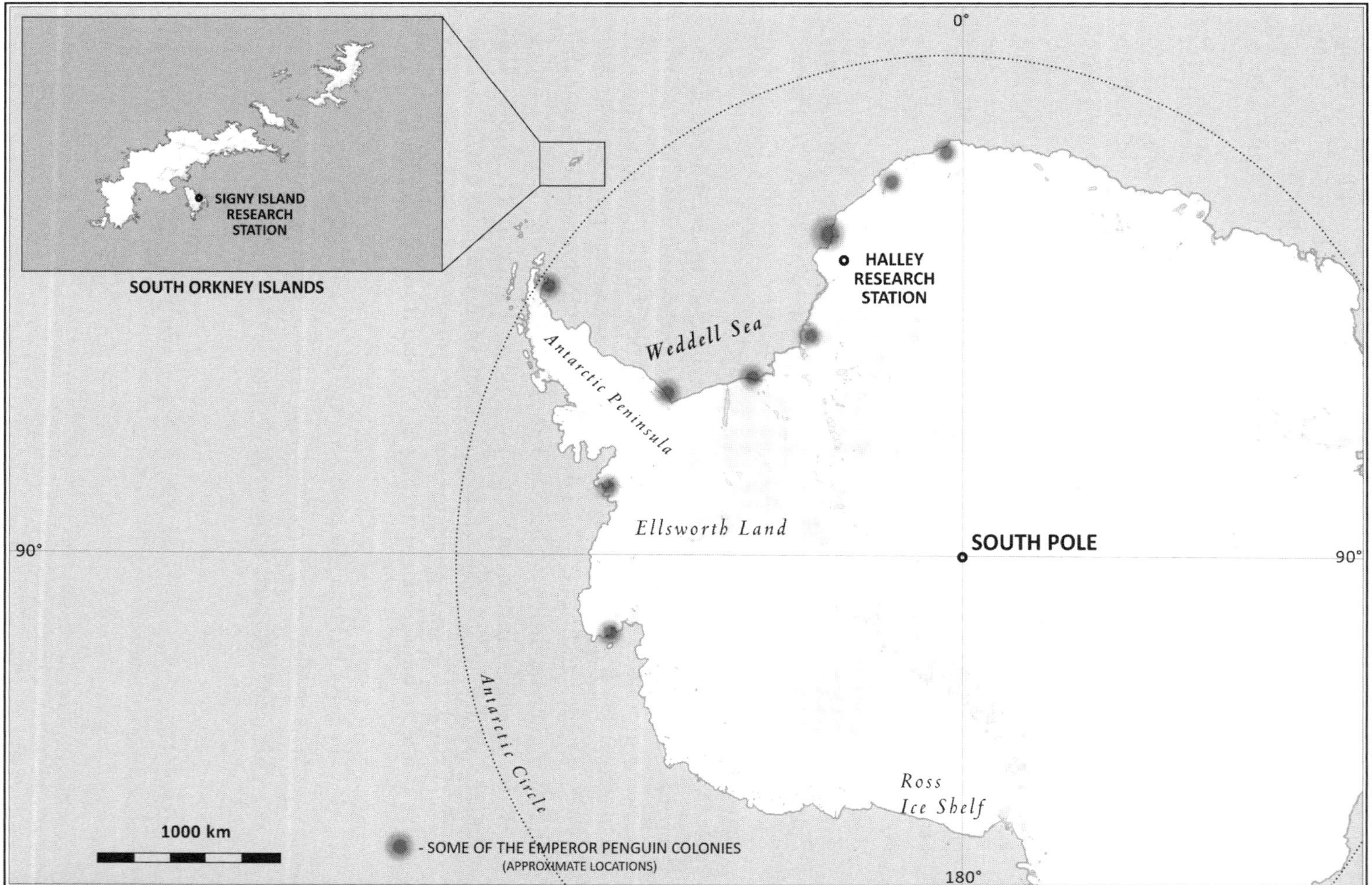

SIGNY ISLAND
RESEARCH
STATION
SOUTH ORKNEY ISLANDS
0°
HALLEY
RESEARCH
STATION
Weddell Sea
Antarctic Peninsula
Ellsworth Land
90°
SOUTH POLE
90°
Antarctic Circle
Ross
Ice Shelf
1000 km
- SOME OF THE EMPEROR PENGUIN COLONIES
(APPROXIMATE LOCATIONS)
180°

Doug Allan

Doug Allan is considering which seal species make the most memorable underwater encounters. Southern fur seals in the seas around the sub-Antarctic islands are among his favourites. He says: 'They start their socialising late in the afternoon, in the shallows, and that's when you can go out and snorkel with them. But when they start treating you like one of the boys, it's a backhanded compliment, because they're quite likely to come up and nip you!' [1]

He is quick to explain that Southern fur seals in the water are generally not aggressive or territorial; they're more likely to deliver a nip out of sheer high spirits. However, he admits that fur seals have sharp teeth, and there can be repercussions: 'There's a painful condition called 'seal finger' which the old sealers and whalers often had. A seal bite can cause a serious infection in the finger joint muscles, that pulls the affected fingers in towards the palm of your hand.' [1]

But when Doug dives into the near-freezing water of the Antarctic, the risk of 'seal finger' is one he's prepared to take. The chance to spend long periods of time with all kinds of marine mammals in their natural environment is a privilege granted to very few people; Doug has embraced those opportunities and made them his life's passion, using his skills as a diver and cameraman to capture footage of rarely-witnessed behaviour. 'Never to see a seal underwater,' he says, 'is like being denied the sight of a bird in the air.' [2]

'Never to see a seal underwater is like being denied the sight of a bird in the air.'

As a boy, Doug was drawn to water. He learned to swim at a young age; at the public swimming baths in his home town of Dunfermline, he would leap in at the deep end and dive down to retrieve bottle tops from the bottom. Inspired by the astonishing underwater films of Jacques Cousteau in the 1950s and 60s, he donned snorkel, mask and fins during family holidays in the Mediterranean. In the clear, warm water, 'I took to snorkelling like a fish,' he remembers. 'I just felt at home in the sea.' [3]

Still in his teens, Doug joined the Fife Sub-Aqua Club and learned to dive with SCUBA equipment. His snorkelling experience gave him confidence, and 'taking on a bottle' allowed him to spend longer underwater. He says: 'I think part of the reason I got into diving was because it was a sport that I enjoyed being good at. It's not competitive but you can recognise a good, elegant diver… they conserve their air, they move quietly through the seascape, they don't kick up the sand. You don't fight with the sea, you learn to work with it.' [3]

Graduating from Stirling University with a degree in biology, Doug was ideally equipped for a research career, but by this time he knew that a laboratory job was not for him. He'd realised that comparatively few marine biologists knew how to dive, while he himself preferred to collect data rather than analyse it. He explains: 'If scientists want an undisturbed sample of a muddy bottom, they have to be able to control their buoyancy perfectly, to hover above it, gently push in the sample tube, pull it out, cork it up, and then take off from the bottom without kicking the mud up. You have to be a good diver to do that.' [1]

Two months after graduating, an advertisement in a diving magazine caught his eye: 'Diver wanted for interesting work in Scottish rivers.' A few weeks later he was working with Bill Abernethy, who is remembered as Britain's last professional pearl fisherman. Doug soon discovered that harvesting pearls from freshwater mussels was 'very hard physically… working against the current, I'd swim over the beds of mussels. From the thousands that were down there, Bill had taught me to pick out the ones that had a particular twist to the shell. Those had the pearls inside.' [1]

Today, the practice is no longer legal, but Doug knows that back in the 1960s and 70s, Bill's work was sustainable because the mussel beds were so extensive and few mussels were taken. He adds that in later years, when it was found that freshwater mussels were in decline, 'Bill had a theory that it was caused by pollutants from agriculture, or salt from the roads being washed into the rivers… freshwater mussels need clean water.' [1]

Doug applied unsuccessfully in the summer of 1975 for a two-and-a-half-year posting as a diver with the British Antarctic Survey (BAS). But in February 1976 he was working as a research assistant in the Red Sea when a telegram arrived: *'Unexpected vacancy on Signy Island for one winter. Prompt reply appreciated.'*

Signy, a BAS research station, is in the South Orkney Islands, about 1,100 miles (1,770 km) south-east of Cape Horn. There was no time for the usual familiarisation procedures; it was, he recalls, 'a case of out of the frying pan, into the fridge. At the end of March, I was dropped off on the base in the Antarctic. I arrived on the BAS ship from Punta Arenas in the afternoon, it left the following morning, and that was it for the nine months of winter! I was right in at the deep end. I didn't know the people there, so it was very much a baptism of fire.' [1]

Living and working in an Antarctic research station means being physically cut off from the outside world during the austral winter. The dozen or so BAS staff communicated with friends and family by Telex messages, which were restricted to a maximum of 200 words per month and relayed to the UK via Stanley in the Falkland Islands. Doug was unfazed by the isolation, focussing instead on his new surroundings. He says: 'When I'm away I like to immerse myself in the environment - I think you can have a much fuller experience if you're not thinking about home too much.' [3]

When he returned to UK after a further four months travelling through South America, Doug found the Antarctic still at the forefront of his thoughts. Within that small, dedicated team he had relished a deep sense of value and belonging, in an environment that he was growing to love. He went down to Signy for another 18 months in October 1978. But at the end of that second contract, just as he was due to come home, fate intervened. Pack ice began to form unseasonably early, making it impossible for the relief ship to reach the base. Attempts to reach the stranded staff had to be abandoned, and they all had to face a further enforced winter at the station.

Doug remembers: 'Two biologists had been on Signy for two years, and now they were stuck there for a third. One person had been on another base for two years, and had come to Signy supposedly for six weeks… he was now there for a year. But toughest of all, there was another biologist who had got married about a month before coming south.' [1]

It was almost as if the Antarctic was testing Doug's commitment, and showing him just how big an impact it could have on his future plans. But he was philosophical: 'I didn't mind. I had no one waiting for me at home. It was a good winter, a great bunch of people. We had solid sea ice for many of the winter months, and did a lot of travelling.' [1] In fact, he was so laid back about the whole situation of being marooned in the Antarctic that instead of boarding the first available ship home, he volunteered to stay behind during the following summer. 'So,' he explains, 'I was there in January 1981, when David Attenborough unexpectedly arrived.' [1]

It was a chance meeting that changed the course of Doug's life. A four-person BBC film crew had hitched a ride on the Royal Navy icebreaker HMS *Endurance* as she made a month-long tour of duty around the Antarctic Peninsula. The ship's itinerary included a visit to Signy to conduct a benthic survey of the bay, and this allowed the BBC crew to come ashore for a couple of days to do some filming. Doug's own stills photography over his summers and winters meant he knew all the best places to go for superb views of glaciers and penguins. He offered to be their guide.

Doug says: 'If I'd met David on a coral reef, it would have been so different because I wouldn't have been anyone special.' [1] But in the Antarctic, with three years' experience behind him, he was ideally placed to help out. In return, he got a glimpse of what life could be like as a wildlife cameraman. With a chuckle, he remembers that 'it was a hopelessly glamorous-sounding profession. After their trip to the Antarctic, they were going to the Galapagos for a few weeks... it seemed to me as if all the elements that were rolled into the cameraperson's job were the things that I'd been building towards - expeditions, underwater work, wildlife, travel, and the excitement and satisfaction of doing something worthwhile.' [1]

This time, when he returned to the UK, Doug had a new direction to follow – filming. But how and where? The British Antarctic Survey offered him another post, this time as the over-wintering Base Commander at Halley Station, located on the Antarctic continent itself. The scientific projects were on pause while the new station was being built, and this gave Doug the opportunity he was looking for. He knew there was an Emperor penguin colony on the coast within 12 miles of Halley. 'Emps nest during the winter. I thought, I'll go to Halley and take a movie camera with me. I'll ask that producer I know at the BBC if he's interested in me acquiring some footage of emps over winter.' [1]

Doug's filming over that winter in 1983 led to another shoot a year later, and when the BBC series *Birds for All Seasons* was aired in 1986, the first 20 minutes of the opening episode consisted entirely of Doug's footage. His film-making career began (almost literally) to snowball. Increasingly, he became recognised as the go-to expert for wildlife filming in cold places. He says: 'It was a great time to get into it. In 1987, the Arctic and the Antarctic were really unexplored film-wise. *Kingdom of the Ice Bear* [1985] was the only big one that had been done

'Expeditions, underwater work, wildlife, travel, and the excitement and satisfaction of doing something worthwhile.'

in the Arctic. I started working with producer Alastair Fothergill, who wanted to do a series on the Antarctic… we pitched an idea to BBC Worldwide, and this resulted in the series *Life in the Freezer*.' [1]

Since then, Doug has travelled the world and worked on many of the BBC's landmark wildlife series, including *The Blue Planet, Planet Earth* and *Frozen Planet*. In the polar regions, in order to get underwater footage of marine animals he'd often be diving in near-freezing water, sometimes swimming beneath the pack ice and around the base of icebergs. The experience can be other-worldly: 'It feels like being in a cathedral lit by suffused light from high windows and the doors closed for peace.' [2]

For some dives, Doug is attached to a safety line, which is paid out by someone on the boat or standing on the ice ; and at times he uses his own judgement about factors such as ocean currents and visibility to decide whether an untethered dive is possible. Ice which is flat above the surface is also flat below, but if it looks jumbled on the surface it is usually much more interesting underwater, forming 'canyons' where he can film animals swimming. The downside is that a dive line can become snagged underwater on rough, crumpled ice.

As a diver, the great thing about filming animals underwater, he says, is that you can 'fly' around them. Understanding an animal's behaviour is key, not just to getting spectacular footage but also to staying safe and being unobtrusive. Doug has learned to recognise the moment when an animal begins to trust his presence. A photograph in his book, *Freeze Frame*, shows him dwarfed by a female humpback whale which is hanging vertically in the water, holding a position where her eyes are angled to get a better look at him. He knows just how fortunate he is: 'Let's say you have a whale coming out of the blue, choosing to spend time with you. You sense how you should swim, even how you should be thinking… so the animal relaxes in your company and carries on with its natural behaviour. It is simultaneously an exciting experience and a massive privilege.' [4]

'You sense how you should swim, even how you should be thinking.'

Respect might engender respect, but sometimes the animals have different ideas. While diving in the Arctic, Doug felt something wrap itself around his thighs. He looked down to see a walrus's head tucked under his arm, its flippers round his legs, having mistaken him for a seal. It was an immediate reaction to hit it on the head with his camera, and it let him go. Later, when speaking to some Inuit friends, he realised what a close call he'd just had: 'They told me that walruses put their lips against the seal's head, suck hard and basically take the seal's brains out. It's a good job that it took hold of me low round my legs, and it didn't get its head near mine!' [3]

Another too-close encounter came in Kong Karls Land, a group of islands off Svalbard, where Doug and his field assistant, Jason Roberts, were filming polar bears. They were washing up in their hut when Doug heard a noise that sounded like someone wiping the window with a squeegee mop. He turned round and saw the wet nose of an inquisitive polar bear rubbing across the glass. Very tentatively, Doug and Jason opened the cabin just enough to direct a burst of pepper spray at their visitor. Luckily, it ran off.

Then there are times when the wildlife subject barely puts in an appearance. In Ladakh, Doug spent 11 weeks looking for snow leopards, also known as 'ghost leopards' because they are so notoriously elusive. One or two had been reported in a remote valley, so, to minimise the risk of disturbance, Doug had a tent and supplies taken in by yak and asked to be left alone for a week, only communicating with his colleagues by radio once a day. Usually, he can tune in to the ambience of the environment, sensing a subtle shift even before an animal becomes visible, but this time he had to admit defeat. 'I only saw one snow leopard in all that time. She walked for 10 minutes across the snow on the other face of the valley, slept for two hours, and then walked out of sight within minutes of waking up. Hardly an Emmy-winning encounter.' [2]

The holy grail of wildlife filming is capturing animal behaviour that has never been seen on camera. For *Frozen Planet* (2011), Doug worked with fellow Scot, Doug Anderson, to film a group of killer whales in Antarctica that were deliberately making waves with their tail flukes to wash Weddell seals off ice floes. Doug first heard about this behaviour, now known as 'wave-washing', while he was with the British Antarctic Survey, but it was another 30 years before he was able to film it.

When TV producers request footage of a blizzard, they don't actually mean a real blizzard. 'In those conditions,' explains Doug, 'you see next to nothing. Contrast is zero, visibility is down to metres, and you might as well film with half a ping-pong ball over the lens for all you'll see.' [2] Conversely, gentler winds and slightly higher temperatures can create ideal 'blizzard' scenarios that will freeze the toes of viewers back home. With the camera low to the ground, Doug films the drifting snow being driven across the ice, backlit by a low sun. His favourite temperature, he says, is -18°C (-0.4°F): any lower, and batteries will die quickly, while sound cables can snap like dry spaghetti.

Doug considers himself privileged to have worked with Inuit people in the Arctic, who helped him to locate narwhal and belugas. He noticed that their attitude towards the animals was always deeply respectful. His guide, Andrew, preferred not to be asked where the belugas would be found that day, 'because it's the belugas who decide where they are going to go, and he didn't want to speak on behalf of the belugas… He did know where to go, but he didn't want to feel as if he was delivering the animals to me.' [1]

The Inuit were very aware, too, of the effects of climate change on their environment. Doug's first dive in the Arctic was in 1988, and at that time, the Inuit who hunted around Admiralty Inlet on Baffin Island knew the exact time each year when a crack would appear in the sea ice, stretching for some 30 miles across the bay. When that crack formed, it was time to retreat from the edge of the ice to safer ground. This knowledge had been passed down from one generation to next, as it was vital to the success of their hunting expeditions. But as little as 10 years later, sea temperatures had changed so much that they were no longer able to predict the cracking of the ice.

Doug shares his own concerns about the Arctic: 'Studies at Churchill on Hudson Bay in Canada have highlighted all too clearly the effects of Arctic warming. Plants on the tundra are flowering earlier. The winter ice is thinner and more broken. The polar bears have a tough time hunting. Their response to the warming climate will be to go further north simply to survive.' [2] Longer ice-free summers mean that pregnant polar bears find it harder to hunt, enter their winter dens hungry, and give birth to fewer cubs.

Reflecting on his early career in biology, Doug says that there was a positive shift in attitudes towards the Earth and conservation during the 1960s and 70s. 'We looked back from the Moon, and that photograph of the Earth just hanging there had a huge impact on people.' [1] He believes the hippies of that era had the right attitudes: 'They had a genuine attachment to the environment, but then the politics of the world changed and consumerism took over. Now, however, people are seeing that the planet's economic models are unsustainable in the long run when they're based on ever-increasing consumption. We're starting to turn our back on it and we so much need that trend to continue.' [1]

Our own awareness of the Earth's natural environments is enhanced by Doug's astounding footage of wild landscapes and seas, and the animals that inhabit them. Sir David Attenborough, with whom Doug has worked on many TV series, has paid tribute to his uncanny understanding of animals, his cheerful endurance of extreme conditions, and his ability to fix all kinds of unexpected problems. Doug sums it up quite simply: 'My heart just happens to be in cold places, particularly in the Arctic and the Antarctic. It's where I feel most alive.' [3]

1 Doug Allan, interview with Jo Woolf (November 2023)

2 Doug Allan, *Freeze Frame: A Wildlife Cameraman's Adventures on Ice* (2012)

3 Doug Allan, *Desert Island Discs*, BBC Radio 4 (July 2014)

4 'Doug Allan: A life capturing the natural world on camera', Graham Fraser, BBC (19th November 2017) https://www.bbc.co.uk/news/uk-scotland-42025005

150°
180°
Arctic Circle
150°
ALASKA
Chukchi Sea
RUSSIA
East Siberian Sea
70°
Beaufort Sea
CANADA
Arctic Ocean
Laptev Sea
80°
120°
2007 approximate extent of ice
90°
NORTH POLE
90°
Kara Sea
Baffin Bay
80°
60°
60°
GREENLAND
Svalbard
Spitsbergen
Barents Sea
Greenland Sea
MURMANSK
TROMSO
70°
500 km
Arctic Circle
ICELAND
30°
0°
30°

Lewis Pugh

o

On board a Russian icebreaker some four days out of Murmansk, Lewis Pugh decides to take a swim in the Arctic Ocean. The water temperature is -1.7°C (28.94°F). Clad only in Speedos and a pair of goggles, he walks to the edge of the ice and dives into the inky blackness.

Lewis has prepared rigorously for this moment. But immediately on entering the water, he knows that something is wrong. His body burns with pain. His goggles freeze to his face. His hands are rigid. After five minutes, he is forced to give up.

Under a hot shower, Lewis stares at his fingers, which have swollen to the size of sausages. There are expressions of shock on the faces of his team. This was merely a trial, in preparation for a planned one-kilometre swim over the North Pole in a few days' time. He has always believed himself to be the best cold-water swimmer in the world. Now he wonders: is he pushing himself beyond the limit of human endurance?

Looking back at his childhood, Lewis Pugh reflects that he has been drawn to the ocean for such a long time that it's probably in his genes. Both his parents served in the Royal Navy, and when Lewis was born they were stationed in Plymouth, Britain's 'Ocean City'. Lewis's mother used to read him tales about navigators and explorers: Christopher Columbus, James Cook, Robert Falcon Scott. Scott's story, played out against the backdrop of the Antarctic, caught Lewis's imagination. Nothing was impossible, Lewis's father told him: things only appear unreachable if you limit your own beliefs.

When his family moved to Cape Town, Lewis gazed out at the ocean through the windows of his new school. From the sea's behaviour, he could work out what the wind was doing: white horses flecked the surface in a brisk south-easterly, while a north-westerly gale hurled big waves against the shore. Antarctica lay some 5,000 kilometres (3,107 miles) to the south: Lewis imagined ice shelves towering above seas that were teeming with penguins and whales. He volunteered as a lifeguard on his local beaches, and at the age of 17 he completed a seven-kilometre swim from Robben Island to Cape Town.

Growing up in South Africa, Lewis noticed the infrastructure of apartheid and he also witnessed its breakdown. He was a law student at the University of Cape Town when Nelson Mandela walked free from prison in 1990, and he remembers the outpouring of euphoria during South Africa's first free democratic election. Looking back, he acknowledges that these two influences - the ever-present ocean and an awareness of human rights - had a long-lasting impact on his life.

As a newly-qualified maritime lawyer, Lewis took a job in London and sought out the ocean in his spare time. Boarding an evening bus to Dover, he would swim for a few hours in the English Channel, snatch some sleep in a guest house and commute back to his desk early next morning. Somehow, he was also finding time to train as an SAS Reservist. He knew he wasn't destined to spend his life in an office.

On impulse, while visiting Norway, Lewis plunged into the Arctic Ocean, and the ice-cold water acted as a catalyst. It was as if he had re-discovered his natural environment. In 2003, he completed the first ever long-distance swim around Norway's North Cape, and in doing so he learned the importance of balancing his speed: move too fast in an effort to stay warm, and cramp would paralyse his muscles, but move too slowly and he risked developing hypothermia. Success confirmed the truth of his father's words: nothing is out of reach, unless you believe it to be.

Nothing is out of reach, unless you believe it to be.

But back among the commuters in London's rush hour, he felt lost. For much of his life, he had studied for a profession that no longer inspired him. A friend had warned him that if he didn't follow his own dreams, he'd end up following someone else's. He knew in his heart that he wanted to be a pioneer swimmer. He listened to his heart, and quit his job.

At first, the challenge itself was validation enough. Over the next few years, Lewis completed long-distance swims in all five oceans of the world, and in 2005 he swam one kilometre at 80° north in Spitsbergen, the farthest north long-distance swim ever recorded at that time. Then, while swimming across Whalers' Bay off Deception Island in the Antarctic, Lewis was horrified to see thousands of whale bones resting on the sea bed. A grim legacy from the whaling industry, they were close enough to touch.

The summer of 2006 brought drought and water rationing to the UK. Lewis was about to swim the length of the Thames when a friend suggested that he use the publicity to highlight the impact of climate change. He pondered this for a while, and felt that he didn't know enough about the science. But he also recalled his happy childhood in South Africa, visiting national parks and being enthralled by the wildlife. He remembered his horror when an oil slick devastated the penguin colonies off Robben Island. He had seen first-hand the evidence of glaciers retreating in the Arctic, and he had touched the whale bones in Whalers' Bay. He realised that protecting the environment was a cause that already resonated with him deeply.

It took 21 days to swim the 325-kilometre (202-mile) length of the Thames; the water level was so low that Lewis had to run the first 40 kilometres (25 miles) from its source in Gloucestershire, and it was so polluted in Oxford that he spent a night in hospital in the middle of the swim. When he passed through London he walked to Downing Street to meet the Prime Minister, Tony Blair, and urged him to take steps to tackle climate change. Lewis recalls: 'It got huge amounts of media attention. I think the thing about the Thames was how easily accessible it was to the media.' [1]

The following summer, he was heading for the North Pole. The icebreaker on which he was travelling had reported open water there seven years previously: this is believed to be the first time that the North Pole has experienced ice-free conditions since the Eocene Period, some 55 million years ago. The evidence of climate change was crystal clear but the science was still difficult for the public to grasp. Lewis wanted to send a direct message: 'Here was something very simple - someone swimming across the North Pole, across an open patch of sea which used to be frozen over.' [1]

'I wasn't swimming in a pair of Speedos for bravado.'

The decision to wear Speedos was deliberate. Lewis explains: 'I wasn't swimming in a pair of Speedos for bravado. I was swimming in a pair of Speedos for two reasons. Number one, I was urging world leaders to be courageous. To make the hard and unpopular decisions necessary now for our future. And secondly, because if you swim in a pair of Speedos you know it's going to be a world story. I wanted all the eyes to be on the Arctic.' [1]

The scientists who had monitored Lewis's previous cold-water swims had noticed that he was capable of anticipatory thermogenesis: his body temperature increased by almost 2°C before he entered the water. This little-understood phenomenon can make a crucial difference between life and death. For several weeks beforehand, he had trained in a glacial lake in Norway, but the fact remained that he had never swum in water as cold as -1.7°C (28.94°F). Sea water freezes at -1.8°C (28.76°F), and water that is on the point of solidifying requires even more effort to move through it. Even with the capacity for thermogenesis, no one could predict the impact that such extreme conditions would have on his body.

As the ship approached the North Pole, Lewis looked back at his trial swim in critical detail. He suspected that he was out on the ice for a fraction too long before getting into the water. Consulting his support team of scientists and trainers, who played a vital role in his mental and physical preparation, he made refinements to his pre-swim procedure. With his entire mind and body, he focused on the prospect of success.

Forty-eight hours later, on 15th July 2007, he dived in at the North Pole. Like last time, his goggles froze to his face, but he told himself that if he continued to think about them, he would die. He could feel his hands freezing, but his body temperature was being monitored and his support team signalled that it was safe to carry on. His route had been marked at 100-metre intervals with national flags, each representing a member of his team, and these positive thoughts spurred him on. By the final stretch, he didn't know how far he'd got. He could only tell himself: *just keep going*. And then he heard them shouting. He had made it. He'd been in the water for 18 minutes and 50 seconds.

A couple of hours later, Lewis stood on the ship's deck with a satellite phone and answered a flood of calls from media outlets all over the world. With his achievement he had sent a clear, direct message and it was now a global story. Everyone wanted to speak to the first person ever to swim across the North Pole. Although he'd barely begun to realise it, his feat was already empowering him to take on a challenge of a different kind.

Crucially, Lewis now felt ready to speak about climate change on the world stage. He embarked on a succession of televised interviews, talk shows, addresses to business leaders and meetings with royalty and politicians. He had risked his life for a cause that he passionately believed in, and he urged action from world leaders in return. And he had no reason to be complacent: in addition to dwindling polar ice, the oceans face other human-created threats from overfishing and pollution. Lewis says: 'I've stood on beaches [in Svalbard] and seen plastic pollution everywhere. I've seen polar bears come out of the sea and climb over this plastic when they want to go and lie down on the tundra.'[1]

In 2013, the United Nations appointed Lewis as their first Patron of the Oceans. His role, as he sees it, is simple: to be a voice for the world's oceans and all the wildlife that lives in them. As an endurance swimmer and a spokesperson, he continues to work tirelessly to fulfil this brief. After a gruelling swim in the Ross Sea, he was instrumental in persuading a number of key nations, including Russia, to set up Marine Protected Areas in the Antarctic. When he completed the first ever swim across the Red Sea, the Egyptian government agreed to create a Marine Protected Area along its coral reefs. In 2023, his swim down the 507-kilometre (315-mile) length of the Hudson River highlighted the importance of clean river water.

Getting reluctant nations to commit to change is a process that demands patience. *Vasbyt*, an Afrikaans word meaning 'bite down hard', conveys Lewis's resilience and determination. Through the Lewis Pugh Foundation, his campaigns have succeeded in protecting 2.2 million square kilometres (0.85 million square miles) of ocean, an area equivalent to the size of Western Europe. His aim is for at least 30% of the world's oceans to be protected by 2030. A significant milestone on the way to this goal is the UN High Seas Treaty of 2023, whose key purposes are to prevent biodiversity loss and to address the effects of climate change in vast areas of ocean that are beyond national jurisdiction.

'There comes a moment,' explains Lewis, 'when you have to make a decision. Are you going to stand up and speak about it or are you going to keep quiet? I decided to speak about the changes I had seen... we have absolutely no right to push animals into extinction.' [2] He knows, too, that the power to make a difference comes from believing in something greater than yourself. From his unique perspective, he will continue to speak out for the protection of natural environments. 'Oceans,' he says, 'are crucial for our very survival - for food, for oxygen, and, I think, for our souls as well.' [1]

[1] Lewis Pugh, interview with Jo Woolf (2019)

[2] Lewis Pugh, interview with Mike Robinson (2022)

PACIFIC
OCEAN
MOLOKAI
Pe'ahi
HAIKU-PAUWELA
Pu'u Kukui
1764m
KAHULUI
LANAI
MAUI
N
Haleakala
3055m
KAHOOLAWE
HAWAII
10 km

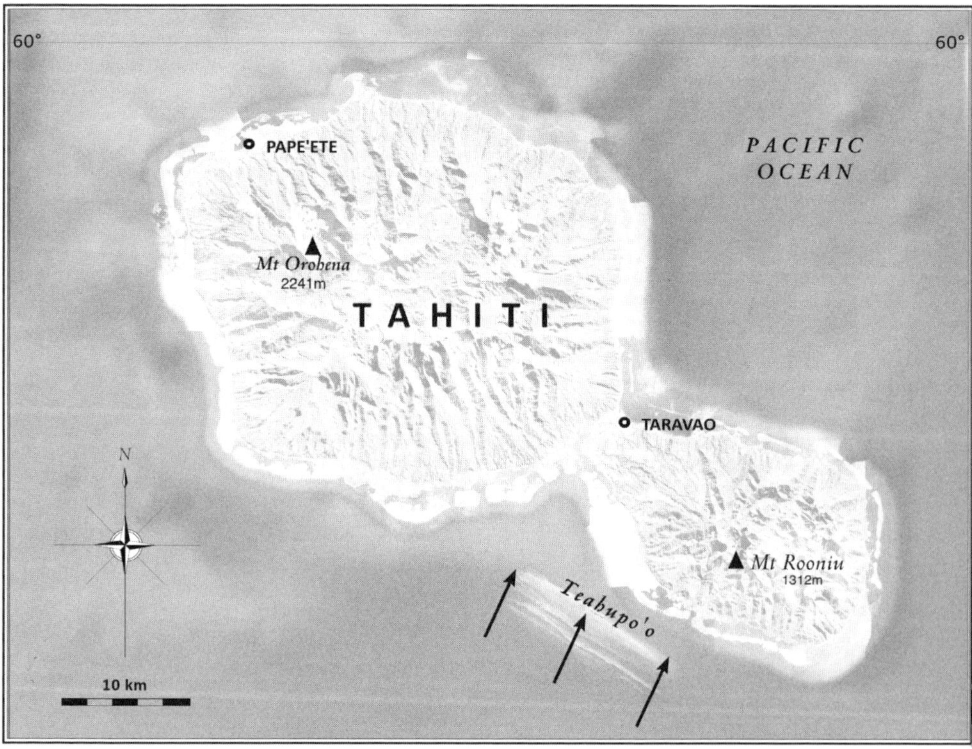
60°
60°
PAPE'ETE
PACIFIC
OCEAN
Mt Orohena
2241m
TAHITI
TARAVAO
N
Mt Rooniu
1312m
Teahupo'o
10 km

Bethany Hamilton

The mesmerising, near-cylindrical barrel waves that break onto the reef at Teahupo'o in Tahiti are born many thousands of miles away in the Roaring Forties, when storms south of Australia and New Zealand send a big swell rolling up towards the Pacific islands. Just off the south-west coast of Tahiti, the steeply-rising contours of the sea bed allow these leviathans, as yet unchallenged, to approach within half a mile of the shore before their 'feet' stumble against this sudden obstacle; then they rise in height and form a powerfully rolling tube of water, flutes of spray lifting from their curling lips, their smooth walls glistening with all the jewel-like colours of the ocean, from deep indigo to translucent turquoise and aquamarine.

In the breathless seconds before these towering waves collapse under their own weight and crash in a wild explosion of foam, human figures on surfboards venture over their brows and slide down their glassy faces, then turn to ride at right-angles to the onshore momentum, half crouching and exquisitely poised as the curving walls close around them, speeding like time-travellers through a vortex that threatens to engulf and consume them.

This is the kind of experience that world-class surfers prize and revere: one that they chase in their waking moments and dream of at night. 'It's like no other feeling,' says Bethany Hamilton. 'You're surrounded by the ocean. You can't explain it fully until you're in there.' [1]

'You're surrounded by the ocean. You can't explain it fully until you're in there.'

For the ancient Polynesians, whose migrations took them for vast distances across the Pacific, the ocean was their mother and their bond. Although their scattered island communities were thousands of miles apart, they were linked by the sea; their *tahuras* - creatures they held sacred - dwelt in the ocean. Using deftly-shaped wooden boards, these early mariners developed wave-riding into a joyful art form, a fluid

Doug Allan with walruses

Doug Allan at an Emperor penguin colony

Lewis Pugh at the North Pole, 2007
Jason Roberts

Lewis Pugh swimming in the Bellingshausen Sea off Antarctica, 2017
Olle Nordell

Bethany Hamilton surfing in Bali
Trevor Murphy

Bertrand Piccard piloting *Solar Impulse*

Solar Impulse

ABB
OLARIMPULSE
UND THE WORLD

Solar Impulse coming in to land in San Francisco
Solar Impulse

Gordon Buchanan with a Bosavi woolly rat
© *Jonny Keeling*

Gordon Buchanan taking a selfie with a wolf, Ellesmere Island
© Patrick Evans

Dinosaur footprint (200 myo *Eubrontes* sp) in the Dolomites, which featured in Chris Packham's *Earth* series. *Bob Gibbons / Alamy Stock Photo*

Mike Robinson on Spidean Mialach above Loch Quoich
Katrine Robinson

dance that responded to every nuance of the wave. Today, in its purest form, surfing is still seen by the Polynesians as an expression of a person's deep connection, not just with the ocean but with the *mana* or energy that permeates all that exists - water, earth, animals, plants and people.

Growing up in Hawaii, Bethany felt this connection without needing to put it into words. She says: 'I think so much of your natural instinct is to be outside. I grew up adventuring - the ocean was my playground. As young teenagers, my friends and I would dive to the ocean floor for seashells, and swim with turtles. I was really immersed in nature, and had a really deep appreciation for it as well. In Hawaii, we have the beautiful ocean and beaches, but we have mountains too - you can go hiking up in the mountains and explore that side of the islands as well.' [2]

Salt water, Bethany once said, is in her bloodline: her parents, Cheri and Tom, had moved to Hawaii for its surf, and both her older brothers, Noah and Timmy, had taken to the water before her. An essential part of a young surfer's skillset is the ability to 'catch' a wave before riding it; Bethany was doing this unaided from the age of seven, and with her family's support she entered local surfing competitions. Her light, flowing style reflected her joy and her oneness with the ocean. As her collection of trophies expanded, she knew that she wanted to be a professional surfer.

Hawaii was - and still is - known for producing surfing legends. In Bethany's generation, a number of talented youngsters were growing up together, coached by professionals and competing against each other with good humour and sportsmanship. Fearless and eager to learn, Bethany was viewed as the one to beat. In July 2003, aged 13, she was placed second in the Open Women's category of the NSSA (National Scholastic Surfing Association) National Championships in California. A few months later, at home in Hawaii, she joined some friends for a pre-dawn paddle in search of good waves.

Then came the phone call that no parent ever wants to receive. Bethany had been attacked by a tiger shark; she was being taken to hospital. Further details were as yet unclear. Meanwhile, paramedics who were attending Bethany radioed ahead to the emergency room. Her left arm was gone.

Despite losing nearly half her blood volume, Bethany survived. The quick actions of her rescuers, and her own calm attitude amid the crisis, had helped save her life. After dealing with the initial trauma, her doctors were optimistic. She was a fighter, and the odds were in her favour.

In a world that was irrevocably changed, Bethany sought hope. She had been raised in a family with a strong Christian faith, and she searched her soul for answers: why had she been chosen to undergo this? What did God have in mind for her life? Although deeply shocked, her parents and brothers were an unfailing source of love. When friends showed their concern, Bethany put on a brave face, but secretly wondered if she would ever surf again.

While still in hospital, Bethany was visited by Mike Coots, a Hawaiian surfer who lost a leg in a shark attack in 1997. Mike had returned to surfing, and assured Bethany that she could do so, too. She knew then that her dreams were still alive. Under medical supervision, she embarked on a challenging programme of rehabilitation and re-learning. Less than a month after the attack, she carried her surfboard down the beach and into the sea.

'Sometimes,' says Bethany, looking back, 'I'm in awe of my 13-year-old self! I guess my love overcame my fear of not being able to do it, or of sharks. That was what helped me to give it a go and figure out how to do it with one arm.' [2] There were serious practical challenges to overcome, such as duck-diving under waves which requires a two-handed grip; to get her started, Bethany's dad added a temporary handle to her board. With absolute determination, Bethany threw herself into re-learning how to balance her weight, and fine-tuning her skills so that she could once again 'dance' with the waves.

In 2004, Bethany entered her first surf competitions; in 2005, she won both the NSSA National Championships (18 and under category) and the O'Neill Island Girl Junior Pro competition. Spectators who might have been fearing the worst were left astounded by her agility and power.

At the same time, Bethany had become a media sensation: her return to surfing was inspiring thousands of people worldwide. Her bestselling book, *Soul Surfer*, and a subsequent film about her life generated a host of public appearances throughout the US and beyond. Although still learning about life herself, she had become a role model for others, and she was keen to help as much as she could. She visited people affected by Hurricane Katrina; in Thailand, she worked with children who were terrified of the sea after the 2004 tsunami; and she visited military personnel who had been wounded in Iraq.

For Bethany, it was a humbling but confusing experience; although she cared deeply about her new commitments, they threatened to dominate her life. Surfing, after all, was the passion that had given her new hope, but in order to share the story of that hope she was being drawn away from the ocean. It took time for her to regain her balance.

Aged 17, she achieved her dream of surfing professionally, and in 2009 she competed for the first time in the ASP* Women's World Tour, at Sunset Beach in Hawaii.

Today, Bethany has a loving supporter in the form of her husband, Adam Dirks, a youth leader and a keen surfer himself. They married in 2013 and have three sons and a daughter. From the outset, Bethany and Adam bonded as a team: both knew that Bethany's heart was in the ocean, and that she still had unfulfilled goals. One of them was to surf in some of the world's most iconic and notoriously challenging waves, including Teahupo'o in Tahiti.

The natural 'mechanics' of Teahupo'o add an extra dimension to its risks: the energy of an approaching wave will suck water out from the shore, leaving behind only a shallow depth of sea. When the wave breaks, any surfers caught up in it can be smashed like jetsam into the sharp coral of the underlying reef; not surprisingly, some choose to wear helmets and body protection. Bethany says: 'It's really fun to grow up feeling comfortable with the ocean, but also having respect for it - knowing your boundaries out there.' [2]

Before surfing at Uluwatu in Bali, Indonesia, Bethany used a trampoline to replicate the lift required to perform airborne twists and spins. With intense focus, she pictured herself making the same movements over and over again. Out on the ocean at Uluwatu, where deep-water channels bring waves crashing beneath spectacular cliffs, she was thrilled to complete a dizzying 'air-360' spin above a breaking wave.

It was back in Hawaii, on the island of Maui, that Bethany challenged herself to surf what is possibly the largest, heaviest and fastest wave in the Pacific. Pe'ahi, also known as 'Jaws' for its ferocity, produces waves over 60 feet high, and they break with a roar that seems to emanate from the very core of the ocean. For world class surfers, the spectacle invokes both excitement and fear, because when things go wrong the consequences are serious. Once trapped inside a wave, a surfer can be dragged with vicious force towards the adjacent cliffs.

When big waves happen at Pe'ahi - which may be only a handful of times every year - crowds gather in an atmosphere of intense anticipation. Ambulances are on call; lifeguards are alerted; safety teams in jet-skis and small boats ride up and down like tiny corks on the majestic swell, ready to snatch up surfers who have hit the water.

Because of the immensity of the breaking waves, most surfers are towed into a prime position by jet-ski, so they are ready to catch a 'set' or group of incoming waves. Rising on the crest of one of these giants, Bethany releases the tow-rope, stands up on her board and cuts a long diagonal line down its face, chased - but never quite caught - by a curving avalanche of water that plunges and crashes behind her, dwarfing her body with its gigantic plume of spray. The ride is fast, elegant, exhilarating. Her smile, immediately afterwards, radiates pure joy.

It is Bethany's belief that you make your own adventure in life; she explains that, when it comes to realising your dreams, the key ingredients are passion, drive, commitment, and the support of the people around you. Like any athlete, she recognises the need to be patient with herself, knowing that one success may only come after a hundred failures. She has always approached life with a 'teachable spirit', embracing the advice of experts from whom she wished to learn; she also acknowledges that her faith is her rock, holding her steady in a sea of challenges.

'I see the beauty and the good that has come out of my situation.'

Looking back, Bethany says: 'I see the beauty and the good that has come out of my situation. I know that God had a plan for my life. The adversity and overcoming had - and still has - a purpose.' [3] With her family and friends, she set up the Beautifully Flawed Foundation, which offers immersive retreats to help young people who have lost limbs. Her team's mission is to inspire others to find hope, as Bethany did, after a life-changing experience.

In 2021, Bethany launched a mentorship programme called the 'Ohana Mother-Daughter Experience, designed to strengthen and enrich the bond between mothers and their young daughters. She says: 'I could not have met the challenges I faced in my life without the deep confidence and resilience and passion that I had. I never needed easy, I just needed possible. The beautiful bond that I have with my mom was part of my foundation, where much of my confidence and resilience came from.' [4] Through the 'Ohana programme, Bethany and her colleagues also focus on improving women's positivity about their bodies, while offering strategies to boost mental and physical health.

Bethany admits that weaving her roles as a professional surfer, mentor and motivational speaker into daily family life can be mentally exhausting. Teamwork is crucial to finding the right balance. When travel is required, Adam and the children will go too: 'I'm thankful,' says Bethany, 'that I can bring my family with me into my world. Being a professional athlete is probably different from having a 'nine-to-five', so to say. I still spend a lot of my day with my family.'[2] And she cares passionately about the difference she is making to other people's lives: 'Living my life so much in the public's eye has been extremely challenging and yet very beautiful at the same time… I am so honored to be able to keep bringing hope to other people.'[5]

Meanwhile, there is always a new opportunity to dance on the waves, and the sound of surf breaking over a Pacific reef will continue to draw her down the beach and into the ocean.

*Association of Surfing Professionals, now the World Surf League (WSL)

1 *Bethany Hamilton: Unstoppable*, documentary directed by Aaron Lieber (2018)

2 Interview with Jo Woolf (May 2020)

3 Bethany Hamilton's website: https://bethanyhamilton.com/

4 Bethany Hamilton, Facebook post, 'Announcing 'Ohana Mother Daughter Experience' (9th December 2021)

5 Bethany Hamilton, Facebook post (28th February 2023)

PALE BLUE DOT

Nature every where speaks to man in a voice that is familiar to his soul.

Alexander von Humboldt, *Personal Narrative of Travels to the Equinoctial Regions of the New Continent during the years 1799–1804*

We can yet make amends, manage our impact, change the direction of our development and once again become a species in harmony with nature. All we require is the will.

Sir David Attenborough, *A Life on Our Planet*

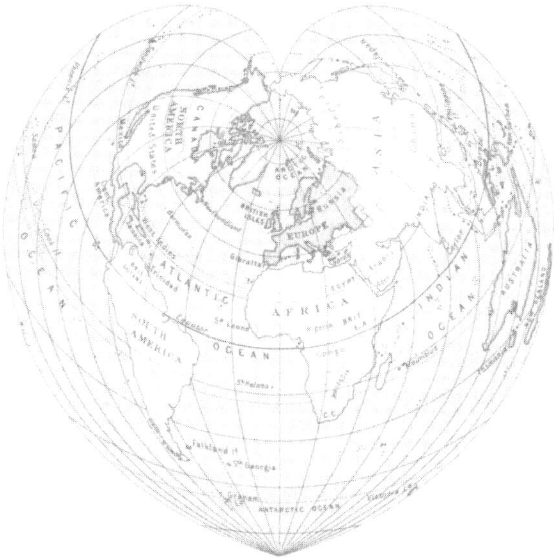

NORTH AMERICA
CANADA
United States
PACIFIC OCEAN
ARCTIC OCEAN
BRITISH ISLES
RUSSIA
EUROPE
ASIA
INDIA
West Indies
Bermudas
Gibraltar
ATLANTIC OCEAN
AFRICA
Trinidad
SOUTH AMERICA
Equator
Congo
Mauritius
INDIAN OCEAN
AUSTRALIA
NEW ZEALAND
St Helena
C.C.
Falkland Is
S.th Georgia
Graham
ANTARCTIC OCEAN

Elizabeth Maruma Mrema

o

The tallest mountain in Africa and the highest free-standing mountain in the world, Mount Kilimanjaro raises its ice-capped head majestically above the hot plains of Tanzania and neighbouring Kenya. This slumbering volcanic giant is ringed with five distinct ecological zones, from the arctic environment that caps its 19,341-foot (5,895-metre) summit, down through alpine desert, semi-alpine heath and moorland, dense rainforest, and cultivated bushland and grassland.

Every year, up to two metres of rain fall on Kilimanjaro's rainforest, where colobus monkeys and hornbills feed in the tall trees and shy bushbuck tread warily through the shadowy tangle of vegetation on the forest floor. More than 1,200 species of vascular plants have been recorded here, many of them endemic to the mountain.

Elizabeth Maruma Mrema grew up in Marangu, a village in Moshi district on the flank of Mount Kilimanjaro, where the sound of running water was ever-present. Rainfall was an almost year-round occurrence; crystal-clear streams tumbled down through the rainforest, nourishing the earth. In homesteads and gardens, the villagers kept cattle and goats, and grew crops such as bananas and coffee. With such an abundance of fresh water, food was plentiful.

But change was on the horizon. Elizabeth explains: 'Unfortunately, as I grew into adulthood, forests and trees kept being slashed for firewood, infrastructure, and development.' [1] Gaps appeared in the rainforest; rainfall levels dropped. The cascading streams, whose soothing voices Elizabeth listened to all night, dwindled and eventually dried up. 'Today,' she says, 'during the hot months, you can see women and children searching for what has now become a very precious resource: water.' [1]

'Forests and trees kept being slashed for firewood, infrastructure, and development.'

It was while she was still at secondary school that Elizabeth noticed these changes happening to the natural environment around her home;

when she progressed to study law at the University of Dar es Salaam, she was already curious to know what had caused them. Her educational path would lead her to the answers, and ultimately it would give her the tools to advocate and implement solutions at an international level.

After graduating in Law with Honours in 1981, Elizabeth continued her academic career with a Master of Law degree at Dalhousie University in Halifax, Nova Scotia, and a Postgraduate Diploma *summa cum laude* in International Relations and Diplomacy from the Centre for Foreign Relations and Diplomacy in Dar es Salaam. As a qualified lawyer, she rose to the position of Senior Legal Counsel for Tanzania's Ministry of Foreign Affairs and International Cooperation, and in the 1980s she was also a part-time lecturer at Tanzania's Centre for Foreign Relations, teaching international law and conference diplomacy.

In 1994, when Elizabeth began working with the United Nations, she brought her legal and diplomatic expertise to bear on global issues relating to the environment. Initially her work focused on supporting countries, particularly those in Africa, in developing and implementing environmental law and in institutional building. At that time, ministries or government departments dedicated to the environment were very few and far between; nor were there overarching frameworks of environmental laws in Africa, as there are today.

The United Nations Conference on Environment and Development, which became known as the Earth Summit, was held in Rio de Janeiro in June 1992. Representatives from 179 countries gathered to discuss humanity's detrimental impact on the Earth's natural environment, and the ways in which it could be mitigated or prevented. The resulting Agenda 21 was a bold programme of action whose recommendations ranged from new methods of education to innovative ways of preserving natural resources.

One of the developments arising from the Earth Summit was the adoption of the Convention on Biological Diversity (CBD), an international legal instrument with the specific aim of assisting cooperating countries in the conservation and sustainable use of the Earth's rich habitats. After numerous leadership positions within the United Nations Environment Programme (UNEP), in 2020 Elizabeth was appointed as the United Nations Assistant Secretary-General (ASG) and Executive Secretary of the Secretariat of the Convention on Biological Diversity.

Already, in 2019, the Intergovernmental Science-Policy Platform on Biodiversity and Ecosystem Services (IPBES) had published a major report that assessed environmental changes over the past five decades. Compiled by experts from 50 countries, and drawing on indigenous and local knowledge, it concluded that around one million animal and plant species are threatened with extinction. More specifically, the report identified five key drivers of biodiversity loss. In descending order, these were our changing use of sea and land; direct exploitation of organisms; climate change; pollution; and the spread of invasive species.

From her childhood in Tanzania, Elizabeth Mrema fully understood the deep connection between these apparently disparate factors; she had witnessed their alarming impact on the rich biodiversity of Kilimanjaro's rainforest. She also knew how crucial it was for the local people to be able to continue to grow crops and keep livestock, following traditional practices to keep their communities healthy and well fed.

Recalling the words of the late Wangari Maathai, the Kenyan activist who was awarded the Nobel Prize for Peace in 2004, Elizabeth observed that it is often the women of a rural community who are the first victims of environmental degradation, because they have the responsibility of walking for hours in search of water and firewood, and for providing food for their families. Often, she says, there is a lack of better alternatives and incentives for women to contribute to the conservation and sustenance of biodiversity.

In a message released on 3rd March 2021, marking World Wildlife Day with a particular focus on forests and livelihoods, Elizabeth explained that habitat loss and degradation, primarily caused by human activity, pose a serious threat to forests and forest species worldwide. An estimated 178 million hectares of forest have been lost in the past 30 years, and forest-dwelling wildlife populations have shrunk on average by more than half since 1970. 'Wildlife loss,' she said, 'has consequences for vital ecological processes that support biodiversity.' [2]

She added: 'The relation between indigenous peoples and local communities and wildlife is tied to their traditional knowledge and customary sustainable use of biodiversity. Recognising their knowledge, innovations and practices along with their traditions, land and values ensures participatory environmental governance for the protection of forests and wildlife.' [2]

In December 2022, when she welcomed international delegates to the UN Biodiversity Conference of the Parties in Montreal (COP15), Elizabeth acknowledged their vast differences in cultures, interests and social and economic development, but emphasised that their common goal was to find solutions to protect ecosystems and to use biodiversity in sustainable ways. Living in harmony with nature was the outcome that they were all seeking, and the resulting decisions would have an impact that lasted not just for the rest of the decade but for the next two millennia and beyond.

'The time has come,' she said, 'to change the trajectory of our planet, and, in turn, the future of humankind. We can no longer continue with a "business as usual" attitude… I challenge every nation and every person to support the adoption of an ambitious, realistic and implementable post-2020 global biodiversity framework; our roadmap to a healthy, sustainable future.'[3] The Conference resulted in the adoption of the Kunming-Montreal Global Biodiversity Framework, a landmark agreement that aims to halt and reverse biodiversity loss, with its 23 targets to be achieved by 2030.

'The time has come to change the trajectory of our planet, and, in turn, the future of humankind.'

The Framework recognises the importance of taking a whole-of-government and whole-of-society approach. In 2023, talking about the potential benefits of the recommended measures to the global economy, Elizabeth pointed out that protecting and restoring nature is good business, but the financial accounts of organisations should reflect the cost of nature loss. For example, it was no longer enough for companies to report to their shareholders on the value gained by cutting down a tree; they should also illustrate the value lost, taking into account the shade that protects from heatwaves, the soil stability provided by roots, the benefits of water retention, and the recreational amenity of the forest.

In her address to the International Geographical Societies Gathering, hosted by RSGS in 2022, Elizabeth said: 'One of our main challenges is reconciling environmental protection with economic prosperity and human wellbeing. Ecosystem degradation already affects the wellbeing of at least 3.2 billion people, which is about 40% of the world's population… Creating and maintaining mechanisms to protect, conserve and sustainably use biodiversity, and ensuring equitable benefit-sharing is not only urgent but crucial for survival on Earth.'[4]

Elizabeth has also spoken about the innumerable benefits of spending time outside in nature. Just by listening to the sounds of birdsong and flowing water in the depths of a forest, or standing on a beach and breathing the fresh air of the ocean, we can reduce our stress levels appreciably; this, in turn, can help to ease a variety of health problems. 'Nature,' she says, 'enriches us in every way, and it simply makes us feel better emotionally as well as physically. Ultimately, I envision an era when we as humans all fully appreciate nature's innumerable contributions to our everyday lives.' [5]

Elizabeth's children grew up in Kenya, where, she says, the sight of elephants and flamingoes comforted them in the same way that she, as a child, had been comforted by the sound of flowing water. But the elephant population has dwindled, and her children wonder what the landscape will look like in 20 years' time: this is just one of many changes to the natural environment that has occurred in recent decades. As Elizabeth pointed out in her message for International Youth Day on 12th August 2022, there are 1.2 billion young people worldwide between the ages of 15 and 24, representing one person in every six. All are likely to be affected in some way by the biodiversity crisis, some of them directly, because they depend on healthy ecosystems for their wellbeing and livelihoods.

It's not surprising, therefore, that young people are expressing their concern - and often their anger - at the apparent unwillingness of governments to take sufficient action in the face of serious threats to the natural environment. In addition, they have a right to expect that their generation, and generations to come, will benefit from clean air, clean water, and sufficient food. Many are taking matters into their own hands, setting up projects to support biodiversity in their communities, and often contributing to scientific research while at the same time using their innovative technological and influencer skills to raise awareness on critical issues. Some are taking their governments to court, pushing them to fulfil commitments made at an international level.

The voices of young people, says Elizabeth, 'are ringing out louder now than ever before and they are crying out on behalf of our planet.' [6] She welcomes their active participation in decision-making processes; they help keep track of governments' commitments on biodiversity, propose solutions, and turn the resulting commitments into much-needed actions.

One way in which young people can express their concerns, share ideas and contribute to worldwide discussions is by joining a youth movement or network. One example is the Global Youth Biodiversity Network, which represents the voice of global youth in negotiations under the Convention on Biological Diversity processes. Elizabeth makes it quite clear that young people's voices are being heard: 'Let us answer the call of youth as they speak out on behalf of the planet they are inheriting. Let us welcome their vision, their input and their passion. Let us allow them room to participate and grow as, together, we build a better tomorrow, today.' [6]

[1] 'Q&A with Elizabeth Mrema', Rhett A Butler, *Mongabay* (20th April 2021) https://news.mongabay.com/2021/04/momentum-is-building-for-a-robust-biodiversity-framework-qa-with-elizabeth-mrema/

[2] Message of Elizabeth Mrema, Executive Secretary of the Convention on Biological Diversity, on World Wildlife Day (2021) https://www.youtube.com/watch?v=i1zNSngnFUw

[3] Elizabeth Mrema, UN Biodiversity Conference (COP15) https://www.cbd.int/conferences/2021-2022

[4] Address by Elizabeth Mrema to International Geographical Societies Gathering hosted by RSGS (April 2022)

[5] Elizabeth Mrema, interviewed by Gabrielle Lipton, *Think Landscape* (4th March 2021) https://thinklandscape.globallandscapesforum.org/50537/cbd-executive-director-on-visions-for-a-new-era-for-nature-and-women/

[6] Elizabeth Maruma Mrema, Message from CBD Executive Secretary, on International Youth Day (2022) https://www.youtube.com/watch?v=7sXDShaEobM

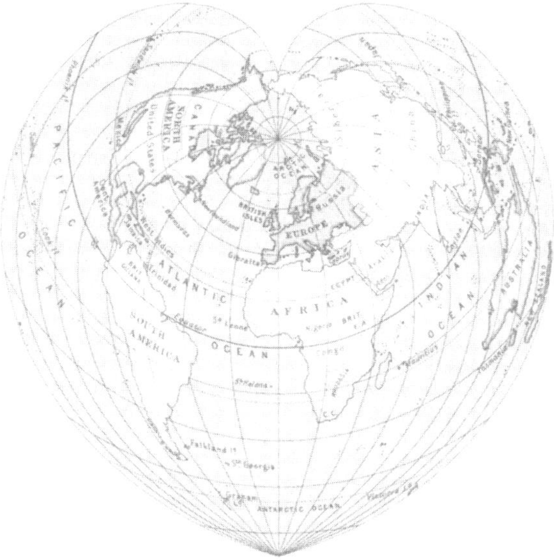
NORTH AMERICA
United States
PACIFIC OCEAN
ARCTIC OCEAN
BRITISH ISLES
RUSSIA
EUROPE
ASIA
INDIA
Gibraltar
West Indies
Trinidad
ATLANTIC OCEAN
AFRICA
EGYPT
Nigeria
Congo
Equator
SOUTH AMERICA
St Helena
INDIAN OCEAN
AUSTRALIA
NEW ZEALAND
Tasmania
Falkland Is
St Georgia
ANTARCTIC OCEAN

Bertrand Piccard and André Borschberg

9th March, 2015. At an airport in Abu Dhabi, an extraordinary aircraft is being readied for take-off. Slender and streamlined, with preposterously long wings, she looks like a giant dragonfly that has spent the night at rest, waiting for the sunrise to warm her body. The ground crew help manoeuvre her into position, an easy task as she is no heavier than an average family car.

At Mission Control, the focus is quiet but intense. Final checks are made; clearance is given. *Solar Impulse 2*, her wings studded with night-lights, skims almost noiselessly down the runway and lifts off into the pre-dawn sky.

Standing on the tarmac, Bertrand Piccard watches the plane leave. He radios the pilot and his voice shakes with emotion. *'Bon vol, André, mon ami, mon frère solaire! Bon vol!'* [1]

The aircraft is soaring now, high above the lights of the city, above the gleaming white domes and minarets of the Sheikh Zayed Grand Mosque. A reply comes back from André Borschberg. *'Merci, Bertrand. Je penserai à toi à chaque minute, et je penserai à l'équipe!'* [1]

It is the beginning of a voyage like no other. Between them, taking turns to be the 'eyes' of the dragonfly in the tiny single-seater cockpit, Piccard and Borschberg are aiming to make the first solar-powered flight around the world. For both of them, *Solar Impulse* embodies a long-held dream that few people believed was possible.

Bertrand Piccard belongs to a family of acclaimed explorers and innovators. As a child, dreaming of adventure was an integral part of everyday life. His grandfather, Auguste Piccard, was a physicist and an aeronaut, the first person to reach the stratosphere in the pressurised capsule which he had invented. Bertrand's father, Jacques, piloted his bathyscaphe into the Mariana Trench, the very deepest part of the Earth's oceans. The young Bertrand chatted to Apollo astronauts and watched their launches. Later, he became a champion hang-glider, while studying medicine and psychiatry at Lausanne University. Aviation both enthralled and fascinated him; in 1999, with Brian Jones, he made the first non-stop flight around the world in a Rozière balloon, *Breitling Orbiter 3*.

Like Bertrand, André Borschberg always dreamed about being a pioneering aviator. In Lausanne, he gained a degree in mechanics and thermodynamics while training as a fighter pilot in the Swiss Air Force. Later, a career in business consultancy and entrepreneurship ran alongside his love of aerobatic displays. From helicopters to commercial airliners, it was said that he could fly any kind of aircraft there was. But could he help design and fly one that didn't yet exist? That was the question asked by Bertrand Piccard, when they met in 2003. André accepted the challenge: he knew that it would be a once-in-a-lifetime adventure.

A mission had taken shape in Bertrand's mind. When *Breitling Orbiter 3* had landed in the Egyptian desert, only 40 kilos of liquid propane remained out of the 3.7 tons that had conveyed the balloon around the world. During the flight, Bertrand had worried that they might run out of fuel. He says: 'I thought, wow, it's wrong to say the sky is the limit. The *fuel* is the limit.' [2]

This revelation gave birth to a new idea. An aeroplane that could fly without fossil fuels, without noise, without causing pollution. Bertrand realised that flying around the world in a solar-powered aeroplane would demonstrate the advantages of clean technology and renewable energy. This, he says, 'was really the moment when all my education, all my poles of interest, all my concerns, could express themselves into one big project. And that was *Solar Impulse*.' [2]

It was an aircraft that the aviation industry had said would be impossible to build. Solar-powered aeroplanes already existed, but they were unable to store energy, so their flight duration was necessarily limited. In order to fly around the world, this new aircraft would have to fly at night as well as in the daytime; it must be lightweight, highly energy-efficient and manoeuvrable. And, of course, it must be safe for a pilot to fly it for several successive days in an unpressurised cabin, often at high altitude.

While Bertrand sought funding and technological support, André led the design team that produced the first prototype in 2009. He says: 'We had to develop the right mindset in order to push the limits of the technologies... When moving into the unknown, it is important to start looking at obstacles as opportunities.' [3] After exhaustive testing, the prototype made successful long-distance flights,

'When moving into the unknown, it is important to start looking at obstacles as opportunities.'

including one that lasted a whole day and night. Just as importantly, it proved to be capable of landing and taking off from commercial airports. After modifications and improvements, *Solar Impulse 2* was unveiled in April 2014.

With a frame of lightweight carbon fibre, a 72-metre (236-foot) wingspan, and a cockpit the size of a telephone box, the aircraft is paved on its upper surface with 17,248 ultra-thin solar cells. Energy gathered from the sun is stored in four batteries, which in turn drive four 13.5 kW motors to turn the propellers, generating an airspeed of up to 50 miles per hour.

As the aircraft's pilots, Bertrand and André had already undergone rigorous physical and mental training. Not only did they have to learn to control the aircraft and deal with potential emergencies, but they had to be prepared to sit in the same position for up to five days and nights, with temperatures in the unpressurised cabin ranging between -20°C (-4°F) and 40°C (104°F). Sleep would have to be taken in 20-minute bursts, relying on an autopilot system to keep the plane on track.

On long flights, *Solar Impulse* would rise to about 28,000 feet (8,534 m) during the day, in order to capture as much energy as possible from the sun; at these altitudes, the pilots had to don an oxygen mask. Then she would descend slowly, powering through the night on stored energy, until sunrise revived her dwindling batteries. Around the clock, a ground crew of 60 people monitored the wellbeing of both pilot and aircraft, and studied detailed weather forecasts in order to plan a safe route. This was, perhaps, the biggest challenge of all, as it was essential that the fragile plane didn't encounter rain or strong winds during its 42,000-kilometre (26,100-mile) journey. Around the world, a string of airfields had been chosen as staging points to allow for maintenance and exchange of pilots. At each one, a custom-made inflatable hangar would welcome the giant bird into the shelter of its own protective wings.

Contemplating the journey, André reflected that his training as a fighter pilot would allow him to evaluate risk and make the right decisions. But jet fighters, as he rightly observed, have the power to fly through turbulence. *Solar Impulse*, on the other hand, reminded him of a leaf being pushed around by the wind: to fly it successfully, you must make nature a partner, not an enemy. Bertrand, who was now a fully qualified pilot, felt that he could also draw on his experience of spending 35 days in a hot air balloon in order to cope with the long flights. He was conscious of the risks; if you have no fear at all, he said, it means you're totally crazy. When you are attempting to do something that no one has done before, your doubts are important. But in order to succeed, somebody has to try.

The departure date of 9th March from Abu Dhabi was timed to avoid the monsoon season as *Solar Impulse* flew eastwards, over India. After 13 hours, André landed in Muscat, Oman, and Bertrand took over on the second leg, flying to Ahmedabad. And so they progressed, to Varansi, then Mandalay in Myanmar, and on to Chongqing in China. At Nanjing, which was to be their last stop before the five- or even six-day flight over the Pacific to Hawaii, every last gram of unnecessary weight was removed from the aircraft. André, who would be in the cockpit for this flight, joked that he had made the biggest contribution by losing four kilos himself. A huge challenge loomed ahead, but everyone was optimistic. And then things started to go wrong.

Predicting the weather for five days in advance over a region of ocean nearly twice the size of Europe was never going to be an easy job. Meteorologists anxiously scanned the charts for a viable hole in a developing frontal system, and found one. André set off, but immediately the forecast changed again, for the worse. There was no longer a window of calm in which to fly. After an intense team discussion, *Solar Impulse* was recalled and an unscheduled landing was made in Nagoya, Japan.

But most of the ground crew were still in China, and the mobile hangar was not ready to shelter *Solar Impulse* from the battering wind and heavy rain that hit Nagoya shortly after she landed. No one knew how much damage it would do to the structure. Until the hangar could be installed, the technicians were battling just to keep her on the ground. Meanwhile, in the control room, André and Bertrand fought to keep their project alive. The role of the engineers was to eliminate as many risks as possible: losing a pilot was a scenario that none of them wanted to witness.

As for the pilots themselves, risks were there to be taken; they were an inherent part of the mission. André urged the team to rise above the difficulties; Bertrand argued that waiting would only increase the chances of failure. A new weather opportunity was seen; André was in the cockpit, ready to go. And then, almost at the point of take-off, the fickle weather forced another last-minute revision. With a bank of eager press photographers waiting on the tarmac, the decision was made to abort.

At 3 am on 28th June, nearly a month after he had landed in Nagoya, André took off again. It was, mostly likely, their final chance of success. A distance of 5,545 miles (8,924 km) lay between him and Hawaii; he was utterly committed to getting there. *Solar Impulse* was still within range of being recalled when the auto-pilot alarm started malfunctioning. André could still sleep, but if the plane encountered turbulence one of the wake-up systems wouldn't work. A fix was found, but in the control room an emergency meeting was held. There was concern that the issue could worsen; ultimately, the decision to continue lay with André. Over the comms system, he assured them that he was confident. By now, he said, he knew the aeroplane well; he had learned what could and couldn't be done. He knew that it had a good source of energy. For him, there was no going back. He set *Solar Impulse* on her morning climb towards the sun.

After the conflicting emotions of that first day, André gazed out at the sunlit sky as *Solar Impulse* floated above an apparently endless ocean. It had not been an easy decision, but now, at least, he had time to adjust and reflect. His wellbeing would depend to a large extent on quality of sleep, and he drew on his long-term practice of meditation in order to quiet his mind.

Meanwhile, the tiny cockpit carried the hopes of thousands: on social media, good wishes were pouring in from followers all over the world. Reading them, André was touched; these beautiful messages buoyed him up for the long and often turbulent flight that lay ahead. Whenever the plane climbed to high altitude, the ground crew monitored his breathing anxiously for signs of hypoxia. But on 3rd July, the long-winged bird with its constellation of night-lights materialised out of the morning sky in Hawaii.

Relief, thankfulness, reunion, quiet celebration. André had made it, and was in good shape. Unfortunately, *Solar Impulse* was not. Batteries had overheated, causing damage that would take a long time to put right. Time was moving on, and so was the season. As André had already observed, with *Solar Impuls*e they must work with nature, not against it. This time, there was no choice. Reluctantly, they postponed the remainder of the voyage.

In April 2016, Bertrand completed the flight across the Pacific, flying from Hawaii to San Francisco. It was a journey of 2,539 miles (4,086 km) and took him three days. En route, he spoke via a live video link to Ban Ki-moon, the United Nations Secretary-General, telling him that *Solar Impuls*e was like a smart grid, collecting and distributing energy with high efficiency. In turn, Ban Ki-moon shared the good news that 175 countries had just signed the UN Climate Change Agreement.

Solar Impulse soared over the Golden Gate Bridge and into San Francisco; then across America to New York. Bertrand was reminded of his father, returning from a month-long mesoscaphe mission to study the Gulf Stream, and receiving a triumphant welcome beneath the

Statue of Liberty. Across the Atlantic, like the pioneering aviator Charles Lindbergh whom Bertrand had met as a child; and then from Seville to Cairo, where André gazed down on the pyramids emerging from the haze.

When *Solar Impulse* took off for the last time, it was Bertrand who was in the cockpit. Before he boarded the plane, he reflected that he'd never been involved in a project that had delivered so many moments of both disappointment and elation. His two-day spell at the controls was far from straightforward: he was sweating uncomfortably at 45°C (113°F) while still wearing the insulated jacket that he'd needed in the freezing air at high altitude. Turbulence stole his sleep, but *Solar Impulse* delivered him safely home. At last, André stood on the tarmac at Abu Dhabi and watched the return of his friend, his *frère solaire*.

26th July, 2016 was an extraordinary day. It marked the conclusion of the first ever round-the-world flight powered entirely by solar energy. By flying for 118 hours between Japan and Hawaii, André had also broken the record for the longest uninterrupted solo flight in an aircraft of any kind.

'I was flying with no fuel, with no noise, no pollution.'

But the flight of *Solar Impulse* was more significant even than that. Although the aircraft itself is not practical as a passenger-carrying vehicle, it is tangible evidence of what can be achieved with clean technology and renewable energy. Bertrand explained: '*Solar Impulse*, for me, was really a symbol of the gap existing between all the current technologies available and the world of the past in which we're still living. I was flying with no fuel, with no noise, no pollution… and I was above the world that still uses thermal engines which are wasting 80% of the fuel you put in the tank. A world with badly-insulated houses, inefficient heating, cooling and lighting systems, old industrial processes, dirty ways to use fossil energies.

'My goal, after flying with *Solar Impulse* around the world, and even with *Breitling Orbiter 3*, was not to say, "The world is beautiful, nature is magical, life is a miracle, we have to protect that." That's not the language of the key decision makers. Their language is about job creation and profitability. So we have to give them that, with the ecological solutions. They need to have profitability and job creation. Ecology is not any more something expensive, boring and sacrificial. It is something exciting. It is something economically profitable. It's something that can bring people together.' [2]

With the establishment of The Solar Impulse Foundation, which offers a wealth of energy-efficient solutions to decision-makers all over the world, Bertrand is already proving that this can be done. Looking ahead, the Climate Impulse project, unveiled by Bertrand in February 2024, aims to undertake a round-the-world flight in a zero-emission, green hydrogen aircraft by 2028.

After the successful completion of the *Solar Impulse* project, André Borschberg co-founded H55, a company that develops and sells electric propulsion technologies to aircraft manufacturers. In 2021, to mark the fifth anniversary of *Solar Impulse*'s epic flight, André and Bertrand shared the cockpit of a small, zero-emission electric aeroplane designed specifically for commercial use. At the time, André predicted that many more aircraft will be converted into electric aeroplanes or newly designed to benefit from electric propulsion technology. In 2024, more steps have been taken towards that reality, with electric-powered or hybrid commercial aircraft in various stages of production.

Working together for many years on the development of *Solar Impulse*, André and Bertrand brought different skills and backgrounds to their partnership; often there were differences of opinion, but both understood that these differences were a unique source of creativity on which they could draw. One thing that they had in common was an absolute refusal to accept that something could not be done, simply because it had never been done in the past. Bertrand expresses this succinctly: 'If everyone tells you that it's possible, that means your dreams are not big enough.' [4]

Accepting the RSGS Mungo Park Medal which was awarded to himself and André Borschberg in 2018, Bertrand Piccard said: '*Solar Impulse* has not only gone around the Earth but it tries to protect the Earth. And I think this is the task of the explorers - it's not just to explore, but it's today to bring this message to key decision leaders and opinion leaders. Because when explorers speak about the world and about the Earth, they are credible because they have seen it. When explorers speak about life, they are credible because they risk their lives. When explorers speak about people, they are credible because they know what is a team. So I believe that we should never waste an opportunity to promote this message of having a better world and showing to the people who rule this world that they have to become explorers also.' [5]

[1] *Point of No Return*, documentary directed by Noel Dockstader and Quinn Kanaly (2017)

[2] Bertrand Piccard, interviewed by Sue Stockdale for *Access to Inspiration* podcast (August 2023)

[3] André Borschberg's website: https://www.andreborschberg.ch/

[4] Bertrand Piccard's website: https://bertrandpiccard.com/

[5] Bertrand Piccard, acceptance speech on presentation of RSGS Mungo Park Medal (2018)

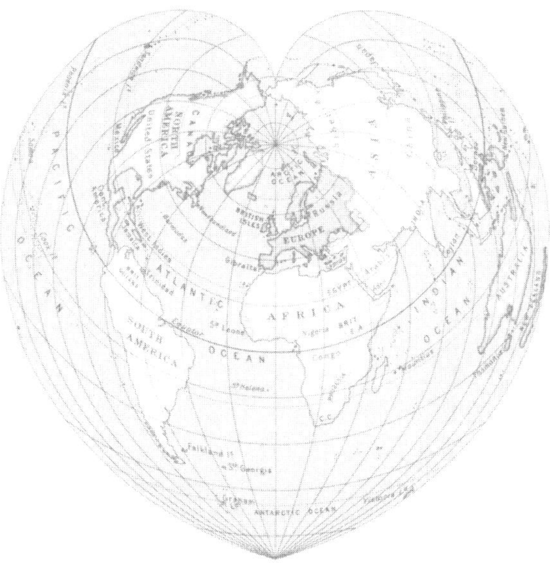
PACIFIC OCEAN
NORTH AMERICA
CANADA
United States
ARCTIC OCEAN
BRITISH ISLES
Russia
EUROPE
ASIA
INDIA
Gibraltar
ATLANTIC OCEAN
West Indies
Trinidad
AFRICA
Egypt
Nigeria
BRIT. E.A.
Congo
C.C.
Equator
St. Leone
SOUTH AMERICA
St. Helena
Falkland Is.
Sth. Georgia
ANTARCTIC OCEAN
INDIAN OCEAN
AUSTRALIA

Chris Packham

o

Chris Packham recalls the first time he climbed out of his bedroom window in order to watch foxes. He was nine years old. Leaving a pillow decoy of himself in bed, he wriggled from the windowsill onto a single-storey roof, slid down to the gutter and dropped to the ground outside his back door.

His secret destination was a den that he had discovered at the edge of a wood. As dusk fell and the last dog walkers headed homewards, he trod quietly and cautiously up a sandy hill and settled down behind a willow bush to wait with barely controlled anticipation.

Nothing happened until it was almost fully dark. Then, all of a sudden, the cubs burst from the bank with a whirl of energy, rolling and bouncing together in boisterous play. Shortly afterwards their mother emerged. Alert and wary of danger, she gazed around until she was satisfied that all was well. Then she walked away into the night, trailing her troupe of tumbling youngsters.

The spell had been cast. Thereafter, on successive evenings, Chris made more surreptitious exits from his bedroom window. He crept ever closer until one night he was just a few yards from the entrance to the den. The position, once chosen, made it impossible to retreat. This time one of the cubs saw him. It approached curiously, fearlessly, until its black whiskers and button nose were just an arm's length from his face.

He was so still that he could hear himself blinking. This fox cub was the most beautiful thing he'd ever seen. After a second it turned and disappeared.

Looking back, Chris admits that, although he wrote polite notes for his parents and pinned them to his door - something along the lines of 'I feel very tired/sick and don't want to miss school tomorrow' - it would have taken chains and bars to keep him in. The voice of the natural world that called him was too loud and too strong. For their part, his parents actually knew about his crepuscular forays and took turns to wait up for the furtive scuffling on the roof that announced his return.

By that time, Chris's obsession with wildlife had long been apparent. Even before he reached school age, diverse forms of insect life were being being ushered into matchboxes and scooped into jars. He probed the ponds on Southampton Common for frogspawn and carried it home, eager to watch it hatch. When he wasn't sleeping, curiosity propelled him outdoors. There was nothing about nature that scared or disgusted him. One or two aspects of human behaviour, on the other hand...

'It is unthinkable,' writes Chris, 'that anyone could maintain an interest in wildlife without formulating a parallel interest in its conservation and the environment.' [1]

Whenever threats to the natural environment are being discussed - in newspapers, TV programmes, public protests - Chris's voice can often be heard urging immediate and decisive action. He is unafraid to challenge government bodies and charitable organisations about issues ranging from fox hunting and badger culling to changes in carbon emissions policies. He says: 'There will never be room for any complacency and contentment in my life as a conservationist and an environmentalist.' [2] In fact, in his memoir, *Fingers in the Sparkle Jar*, he writes that contentment must be avoided at all costs, and the fuel to accomplish that is dissatisfaction. 'My approach,' he explains, 'has always been forthright and relatively ferocious.' [2]

'My approach has always been forthright and relatively ferocious.'

This staunch and unflinching attitude has been reached only after wading through a morass of emotional pain in his early life. In the 1970s, very few people had even heard of Asperger's. Chris's diagnosis only happened when he was in his forties, and he has since spoken candidly about the bullying he experienced at school, and his youthful anguish and self-loathing. As a boy and then a teenager, watching wildlife not only fed his passion, it alleviated his anxiety and allowed him to feel more in tune with the world.

Chris's sense of rejection and isolation sparked an urge to rebel against the prejudices of authority, and this found an outlet in the punk rock movement of the late seventies and eighties. For a while, he played in a band himself. The music helped, he says, 'because it was full of energy, it was robust and very angry, and I was angry at that point.' Now, he adds, 'I've learned to channel that anger into creative things. Don't waste that emotion - you've got to use that reaction to make it right.' [3]

Graduating from Southampton University with a degree in zoology,

Chris worked first as a wildlife cameraman and then as a children's TV presenter. His wide-ranging knowledge, together with a talent for communication, earned him a reputation as something of a wildlife guru; since then, he has become a respected host of BBC's *Springwatch, Autumnwatch* and *Winterwatch*, and a presenter of wildlife documentaries. Rebellion is still simmering beneath the surface, but now it fuels his determination to protect the Earth's natural environment.

'Short-termism,' he says, 'is our greatest problem. Short-term politics, short-term views from people who go on about the fact that they're resistant to change because that would compromise them in some way.' [2] Climate change, he concedes, can be difficult for people in the UK to perceive as tangible in comparison to other parts of the world, and as a consequence there's a tendency to believe that it's not happening yet, or it's only happening elsewhere. 'Well, that's not the case. It is happening here, and it's doing untold damage to our environment, and that means, in the future, our agriculture, our standard of living, quality of life, and everything else.' [2]

By contrast, Chris believes that young people are a lot less short-sighted: 'They've realised that their future is critically in peril, and rather than pull down the blinkers and carry on regardless, they're saying we don't have that option.' [2] And while he acknowledges that everyone individually can take valuable steps in the right direction, he says that if we really want to make an impact, we must press our governments to combat the broader issues with more speed and urgency.

In Scotland, one of those broader issues is the oil industry. Chris says: 'We know that we can't stick with a carbon-based fuel system so we should stop looking for more oil and invest in a transition from petrochemicals through to sustainable energies.' [2] He acknowledges that the oil industry has provided generations of people with jobs, but the transition is still essential in the face of global temperature rises and pollution. And that change, he says, 'should be happening now, like this afternoon, not at some stage in the future... if we carry on regardless, there won't be any jobs, there won't be any education, because there won't be any hope.' [2]

'Vested interests get in the way, but it just makes us fight harder. Ultimately, we'll win.'

On moorland estates, Chris has repeatedly called for an end to the illegal persecution of raptors, and for driven grouse shooting to be replaced with a more sustainable alternative such as walked-up shooting. Keen to protect Britain's last surviving remnants of ancient woodland, he has criticised developments such as the HS2 rail link which threatens to deplete them further; and around the coast he has drawn attention to the relatively unseen environmental impact of open-cage salmon farming. He explains that elsewhere in Europe and in Canada, salmon farming still generates revenue and jobs but is practised a way that is far less damaging. 'Vested interests,' he says, 'get in the way, but it just makes us fight harder. Ultimately, we'll win.' [2]

Working with organisations such as Birdlife International and the Committee Against Bird Slaughter, Chris has investigated and filmed the seasonal shooting of millions of birds in Mediterranean countries. Every spring and autumn, migrants are either trapped or shot, and the activity is viewed by participants as a traditional sport. The word 'tradition', says Chris, 'seems to want to justify something that doesn't have contemporary relevance. Slavery was a tradition for quite a long time, so was burning witches. It doesn't mean that we continue them - when we realise that they're ethically bankrupt, we stop it.' [2] While filming in Malta, attempts were made to intimidate him, but he remained unmoved: crucially, he had video footage to prove that laws protecting birdlife were being ignored.

Chris's words often provoke strong reactions. He has received several threats to his personal safety. Anonymous protesters have made sickening gestures by leaving dead crows and a dead badger outside his home. In the post, he received a block of wood on which was carved - badly, he pointed out - a penis and testicles. He accused the sender of lacking the balls to deliver their message in person and then auctioned the item on eBay, donating the proceeds to the campaign Wild Justice. These responses, he says, prove that his words are striking home and forcing opponents to lash out. They neither worry nor annoy him. He will plough on regardless.

During the first Covid lockdown of 2020, Chris was inspired by birdsong and some celandine flowers to log into his Twitter account and share the small but wonderful signs of spring that he was witnessing in his corner of the New Forest. That impulse quickly mushroomed into a series of hugely popular videos hosted by himself and his stepdaughter, Megan McCubbin, who is a zoologist and conservationist; both of them were thrilled to witness the joy of people who were discovering wildlife on their own doorsteps, all over the world.

It was a triumph for social media, but Chris himself is unlikely to succumb to the addiction of scrolling. When he and Megan were sofa guests on the Channel 4 show *Gogglebox*, and they heard a judge in a chefs' competition ask if a dish was 'Instagrammable', Chris was comically appalled, saying that if humanity was a hand grenade, he would be tempted to pull the pin out.

In 2023, when he presented the BBC series *Earth*, Chris was dealing with environmental changes on a truly colossal scale. In the first episode, entitled 'Inferno', he explained that at the end of the Permian era (about 252 million years ago), the Earth's crust ripped apart and volcanic material was blasted upwards to a height of six miles. Ash fell like snow, and toxic gases were released into the atmosphere, causing mass extinctions. Some figures suggest that 70% of land vertebrates and 96% of marine organisms were wiped out. Then, in the geological blink of an eye, came a deluge of rain that affected the Earth's climate for another two million years.

Epochs of such vast duration demand an almost super-human ability to comprehend. By contrast, we can also glimpse the most fleeting moments of individual lives: in the Dolomites, Chris was delighted to place his hand in the three toes of a dinosaur's footprint, perfectly preserved in a rock face. More to the point, amid all this upheaval and devastation, life persisted and eventually flourished. In a way, Chris explains, we are all survivors of those cataclysmic events. Every living organism has an ancient ancestor, perhaps millions of generations back, that survived and then prospered in the aftermath. Preserved in the structure of the Earth itself is evidence of the fragility and also the tenacity of life.

That tenacity, that extraordinary power to survive, is a powerful source of hope. But, as Chris points out, it also gives rise to a question. What would happen if we didn't address the climate and biodiversity crisis, what would happen if we burned all the fossil fuels and made the planet into a hell-scape? The answer is that although we would perish, life itself would eventually bounce back in all its diversity and beauty. In that case, he asks, should we bother to take any action at all? It's hard to dispute his answer. The extinction event of the late Permian was part of a planetary process, a natural volcanic eruption. This time, we are tipping the balance ourselves. Do we really want that kind of extinction on our own conscience?

But there is hope here too. Despite seeing so much of the natural world destroyed in his own lifetime, and despite the evidence of human-induced climate change, Chris still has faith in humanity. For all of our flaws, he says, we are still a remarkable species, which has evolved from standing and walking on two legs to drawing beautiful animals on cave walls to building spacecraft and flying to the Moon. We are resourceful, intelligent and creative. We can conjure amazing things out of our own consciousness. Very soon, he predicts, we shall have to learn to put our Earth first.

Chris speaks his mind with conviction and clarity. However, it is easy to forget that behind his words lies a sensitive person with acute powers of perception. When asked about the individuals who particularly inspire him, he gives credit first to the people who supported him through a difficult period in his mid-twenties; he feels indebted to them, and for this reason he is keen to invest in helping young people, particularly those who may be going through a tough time themselves.

'We have a conscience and if we act upon it as individuals we can make a difference.'

He admires anyone who has the courage to stand up for their beliefs: 'I like people who get things done - people who don't care if they upset the apple cart, because they know that it's the right thing to do at that point in time.' In his memoir, he writes about the excitement of a struggle against the odds, and of being teased - only temporarily - by the appalling notion of giving up. 'Self empowerment,' he says, 'is the key to success. We have a conscience and if we act upon it as individuals we can make a difference.' [1]

Most recently, Chris has called on the pop singer Taylor Swift to show that she cares about the future of the Earth - and of her immense cohort of fans - by selling her private jets. He has written to TV hosts Ant and Dec, asking them to stop using live animals in their reality show *I'm a Celebrity... Get Me Out of Here!* And in December 2023 he filed a High Court legal challenge against the British government in response to their decision to delay certain actions laid down in the Carbon Budget Delivery Plan, including the phasing out of petrol and diesel cars.

During *Springwatch*, Chris has often dropped words and phrases from punk rock songs into his live commentary on wildlife. In June 2024, when he addressed the 60,000 people who took part in the 'Restore Nature Now' march in London, he wove them into his call for action: 'Since Frankie Goes to Hollywood told us to relax, 600 million birds across Europe have vanished. Since The Sex Pistols pointed out that it was anarchy in the UK, 80% of our butterflies have gone... Now is the time to stand up and start fighting for all of the life that we love so much… to finally find the courage for the biggest battles, not the easy wins.' [4]

Chris admits that sometimes the issues seem too big, too complex and too uncomfortable for us even to consider, let alone resolve. As we struggle with our response to the gloomy figures and scenarios presented to us, our emotions require new labels: eco-anxiety, climate grief. 'Many of us,' he writes, 'are fed up, irritated and overwhelmed. I know I am.' [5]

But he overrides this response with positivity. He describes himself as the sort of person who counts rainbows. 'You mustn't miss a moonrise,' he says. [3] Compared to the bewildering million-year cycles of Earth's violent past, our tenure here is short, and nature is dazzling, complex and wonderful. 'You've got to be a sponge, you've got to suck every single last bit of that up.' [3] And he certainly embodies that ethos. Whenever he encounters any form of wildlife, his fascination is absolute. 'Now is the time to love life,' he told nature supporters in Parliament Square. [4] He is a wild child with a rebel fist and a compulsive need to shout above the noise.

1 Chris Packham's website https://www.chrispackham.co.uk/chris-packhams-biography

2 Interview with Jo Woolf (December 2019)

3 Chris Packham, *The Penguin Podcast with Richard E Grant* (2016)

4 Chris Packham, Instagram story (June 2024)

5 Chris Packham and Megan McCubbin, *Back to Nature* (2020)

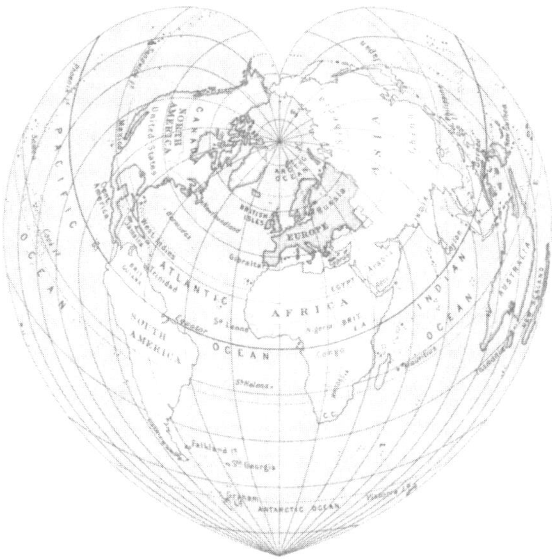
NORTH AMERICA
CANADA
United States
Mexico
PACIFIC OCEAN
BRITISH ISLES
EUROPE
Russia
ARCTIC OCEAN
ASIA
India
Arabia
Egypt
Gibraltar
West Indies
Jamaica
Trinidad
ATLANTIC OCEAN
AFRICA
Nigeria
Congo
St Helena
Equator
SOUTH AMERICA
INDIAN OCEAN
Ceylon
Mauritius
AUSTRALIA
NEW ZEALAND
Tasmania
Falkland Is
St Georgia
Graham
ANTARCTIC OCEAN

Polly Higgins and Jojo Mehta

As a child, Jojo Mehta always thought she was going to be a writer. Her mother, a singer, songwriter and poet, drew her inspiration from the Earth: on country walks with her children she would show them all kinds of wild flowers, teaching them their names and inspiring them with a deep appreciation of nature.

An avid reader, Jojo studied languages at university. After graduating in French and Italian language and literature, she took a Masters in social anthropology. 'Those studies,' she said, helped me to have a kind of empathy with other cultures, with being curious about the world.' [1]

Having spent her twenties travelling and living in different parts of the world, Jojo returned to England to settle down and raise a family. It was then that she heard a relatively new word being used in relation to the natural environment: fracking. This term, used to describe the process of hydraulic fracturing in order to extract oil or gas from bedrock, had hit the headlines in Britain in 2011, when a fracking site in Lancashire triggered two minor earthquakes. Fracking was already mired in controversy, with its attendant risks of pollution, damage to the environment and harm to public health.

'If they're poisoning the ground, surely they understand they're poisoning themselves.'

Jojo remembers: 'I was talking about it with my family and friends, and my daughter burst into tears. She said, "Mummy, I don't understand. If they're poisoning the ground, surely they understand they're poisoning themselves. They're going to die. You've got to call them and tell them to stop." And I said to her, "I don't know what difference it's going to make, me calling one of these big companies up." She said, "There must be somebody you can talk to. Can't you talk to the voting man?"' [1]

During a recent political ballot, Jojo had explained the election process in simple terms to her young children. Her daughter's reaction made her stop and think: what could she do about it? Of course, she was seriously worried about the harm that was being done to the environment, but would her voice actually be heard? While still in doubt, she made an appointment with the 'voting man' - her local MP - and found herself explaining her concerns at his next surgery. He was, she remembers, 'doing that classic politician thing that they do, of side-stepping everything that you ask, of reassuring but not quite answering.' [1]

Jojo came out of the meeting frustrated, but absolutely determined that this wouldn't happen again. In future, she would be better informed; in future, she wouldn't be caught out or fobbed off. She says: 'That's when I guess you could say I put my boots on the ground.' [1]

First, Jojo took the time to inform herself about the facts, and then she started distributing leaflets, giving talks, and helping to organise demonstrations. Fracking, as it turned out, was eventually halted in the UK after widespread public outcry. Meanwhile, thanks to her daughter's plea for action, Jojo had reinforced her own affinity with the Earth. She had found her voice, and she was using it to demand protection of the natural environment, not just in Britain but around the world.

It wasn't long before people were saying to Jojo that she really ought to meet Polly Higgins. Polly, a lawyer, was researching a potential legal case against fracking. 'It was,' recalls Jojo, 'one of those kindred spirit moments... within half an hour of speaking to each other, it was clear that we were very much birds of a feather.' [1]

Like Jojo, Polly's mother was an artist. Around the kitchen table at their family home in the Campsie Fells, north of Glasgow, Polly's father, a meteorologist, often discussed the evidence of climate change that he could see emerging from his work. Polly had an inherent sense of justice, and she developed a strongly-held belief that the world needed to be cared for. She gained degrees in cultural history and decorative arts from the Universities of Aberdeen and Glasgow respectively, and she was working for an art dealer in London when she felt an overwhelming urge to switch careers and become a lawyer.

To her friends and family, this was a puzzling decision: Polly herself admitted that she was uncomfortable with public speaking. But her strong sense of justice was the driving force. At school, despite her shyness, she had taken decisive action against bullies who targeted herself and others. So she trusted her intuition and enrolled as a law student at City University in London. She progressed to the Inns of Court School of Law, and was called to the Bar in 1998.

One afternoon, while practising as a barrister in London's Royal Courts of Justice, Polly had a moment of revelation. She was representing a man who had been badly injured in the workplace. During the hearing, she glanced out of the window and a startling thought occurred to her. The Earth, just like her client, had been badly injured by human activity, and yet nothing was being done about it. To Polly, it seemed as if the Earth was in need of a good lawyer.

'That thought,' Polly recalled afterwards, 'didn't leave me alone... I thought, as a lawyer, what are the tools that I need to represent the Earth in court? And what I realised was that they didn't exist.' [2]

'As a lawyer, what are the tools that I need to represent the Earth in court? And what I realised was that they didn't exist.'

It was a conundrum. All humans have a right to life. But Polly wanted to know: what if the Earth had a right to life as well? She consulted her colleagues in the legal profession and they told her that she was crazy. Of course, the Earth didn't have rights. And meanwhile, there was a whole body of environmental law to deal with individual disputes about damage to the natural world. But to Polly, it was clear that existing environmental laws were ineffective. In the Amazon, for example, large-scale destruction of the rainforest was happening on an ever-increasing scale. Existing laws were powerless to stop it.

It seemed as if Polly was alone in believing that the Earth has a right to life. But she wasn't. When she looked around, she discovered that millions of people shared her point of view. Many were indigenous people, whose connection with nature was long-standing and deeply respectful. To Polly, it felt as if the 'civilised world' had forgotten something that underpins our very existence.

The concept of the Earth as a living organism is nothing new: it can be traced right back to early human cultures. Earth-deities, often in the form of a mother-goddess, occur in ancient myths and belief systems from Polynesia to West Africa and from northern Europe to South-east Asia. These deities are often linked with fertility and the natural cycle of the seasons: they embody and influence all the elements that support life.

There are, explained Polly, two different ways of viewing the Earth. One is to see it as an inert object, a commodity to be valued and exploited for our own benefit. But when we consider the Earth as a living being, we stand in a very different place, and this brings about a dramatic shift in our awareness. Once we see ourselves as trustees, as guardians of the Earth, we can take responsibility for our actions, and we start to consider the welfare of future generations.

Polly could see that the most effective way to stop the abuse was to legislate against it. As a barrister, she fully understood the laws that govern human rights. The Rome Statute of the International Criminal Court, which came into force in 2002, defines four core international crimes: genocide, war crimes, crimes of aggression and crimes against humanity. All are based on the fundamental human right to life. But what about other life forms, other species, other inhabitants of the Earth? Polly was convinced that there should be an additional international crime: that of ecocide.

The word 'ecocide' was first used in the 1970s by the American biologist Professor Arthur W Galston to describe the use of the herbicide Agent Orange by US forces during the Vietnam War. Decades later, thanks to Polly's efforts, the concept of ecocide re-emerged onto the global stage. Ironically, her research had revealed an uncomfortable fact: during the drafting of a precursor to the Rome Statute, ecocide was included, but the clause was removed after objections by a number of oil-producing countries, and by the time the Rome Statute was negotiated the clause was not included. From that moment, Polly's mission was to advocate for what she saw as the 'missing crime' of ecocide to be reinstated.

In 2010, Polly proposed a definition of ecocide for potential consideration by the United Nations International Law Commission. Ecocide, she argued, was 'the extensive destruction, damage to or loss of ecosystem(s) of a given territory, whether by human agency or by other causes, to such an extent that peaceful enjoyment by the inhabitants of that territory has been severely diminished.' [2] The key word here was 'inhabitants': Polly was thinking not just of human beings, but of all other living organisms.

And there were other elements to consider, too. Pollution, for example, may not be immediately apparent, but it could affect the wellbeing of future generations. Natural disasters such as tsunamis and rising sea levels can cause ecosystems to collapse. Under the proposed legislation all nations, regardless of their geographical location, would have a duty of care.

In many cases of ecocide, Polly was aware that impactful decisions were being made by heads of state and chief executives of corporations seeking to maximise their shareholders' profits. A corporation, she acknowledged, has a legal duty to put profit first. That principle might have served us in the past, but we didn't consider the consequences to the natural environment. There should now be a 'think before you act' provision, placing greater responsibility on those in a position of control.

In September 2011, Polly set up a mock ecocide trial in the UK Supreme Court in London. This unprecedented event involved a judge, a jury, and teams of highly regarded barristers for prosecution and defence. The case focused on a fictionalised environmental disaster, and the ensuing courtroom proceedings allowed the implications of the proposed law to be examined, revealing how it would work in practice.

When Jojo met Polly, she had already heard about the mock trial. She says, 'I remember thinking, what a fantastic campaign!' [1] Polly and Jojo worked in parallel for a couple of years, with Jojo advising Polly on communications and Polly contributing her legal expertise. In 2017, they set up a global campaign now known as Stop Ecocide International. Polly, explains Jojo, already had a large international following: 'What we ended up doing was combining the campaign side and the legal side and creating a public campaign to support moving this law forward.' [1] Soon they were attracting supporters from all over the world, including influential public figures.

Jojo says: 'I think people don't always remember that criminal law is essentially protective law... what it's really there for is to prevent the crimes from happening.' [1] Destruction of ecosystems is still proceeding at an alarming rate, indicating that there is not enough of a deterrent or incentive to change our behaviour. In addition, humans are creatures of habit, and we don't change our ways unless we have to. But with ecocide as a criminal law, says Jojo, 'key decision makers would be pausing with their pens over the dotted line, going, hang on, is this going to put me in a position where I could be criminally liable? Because I know this could create severe harm.' [1]

Sadly, Polly passed away in 2019, at the age of 50. Jojo remembers her friend with affection and respect: 'Everybody who met Polly almost fell in love with her. She had an inspiring visionary quality that was uncompromising and yet incredibly compassionate and non-judgmental.' [3]

Afterwards, Jojo initially feared that their campaign would lose momentum, but in fact many people stepped forward, eager to help. After several years of hard work, Jojo says that 'what we have now is really quite legitimately described as a global movement, because of all those connections and bringing all of that collaborative energy together.' [1] Support has been swelled by other campaigns around the world, including those led and promoted by young people.

Stop Ecocide International continues to pursue its mission to see ecocide accepted as a crime by the International Criminal Court. In June 2021 a legal definition of ecocide, commissioned by the Stop Ecocide Foundation and drafted by an Independent Expert Panel of top lawyers from around the world, was launched to the international media. Ecocide, it stated, means 'unlawful or wanton acts committed with knowledge that there is a substantial likelihood of severe and either widespread or long-term damage to the environment being caused by those acts.' [4]

On 9th September 2024, the oceanic state of Vanuatu submitted a formal proposal to amend the Rome Statute to include the crime of ecocide using this definition, and this proposal was co-sponsored by two more Pacific island nations: Fiji and Samoa. Ralph Regenvanu, Special Envoy for Climate Change and Environment for the Republic of Vanuatu, explained that the island's economy was being devastated by climate change and environmental loss, causing the submergence of territory and threatening livelihoods.

Less than two months later, the Democratic Republic of Congo became the first African nation to formally endorse the creation of an international crime of ecocide. The DRC faces serious environmental threats from deforestation, illegal mining and climate change. By November 2024, proposals for bills to introduce national ecocide legislation were progressing in the Netherlands, Mexico, Brazil, Peru, Italy and Scotland; ecocide legislation has already been introduced in Belgium.

Back in March 2020, when Jojo recorded a Mother's Day message for Stop Ecocide International, the world was reeling from the impact of Covid lockdowns and parents were struggling to home-school their children. But alongside this crisis, we were hearing about devastating fires in Australia and the Amazon, and floods in Indonesia.

Jojo admitted that the situation had made her think about why her children were in school at all: was it to train to get a job, perpetuating a consumer culture that fills the seas with plastic and the sky with fumes? 'As mothers,' she said, 'our strongest and deepest urge, in common with most living creatures on this Earth, is to protect our children. We'd run into a burning house for them. And right now, our collective house is burning.' [5]

> ***'It's our generation, it's the incumbent generation that need to wake up and do something very quickly.'***

Jojo is clear that we cannot expect our children to deal with environmental crises when they grow up. On the contrary, they should be demanding immediate action from us: 'It's our generation, it's the incumbent generation that need to wake up and do something very quickly.'[1] But she can also see positive aspects in our predicament. After all, she says, 'how many generations get to affect the future of their entire species with what they do? I believe we are living through one of those times right now, and that is an incredible opportunity.'[1]

[1] Jojo Mehta, interview for RSGS Inspiring People Talks (2022)

[2] 'Ecocide - the fifth crime against peace', TedX talk by Polly Higgins, Stop Ecocide International (2012) https://www.youtube.com/watch?v=Qg9eWtmBMIQ

[3] 'The legacy of Polly Higgins', Jojo Mehta, Stop Ecocide International (2019) https://www.youtube.com/watch?v=VNuUjJOxprA

[4] Legal definition of ecocide, Stop Ecocide International https://www.stopecocide.earth/legal-definition

[5] 'A Mother's Thoughts', Jojo Mehta, Stop Ecocide International (2020) https://www.youtube.com/watch?v=vVK1fbrKfSs

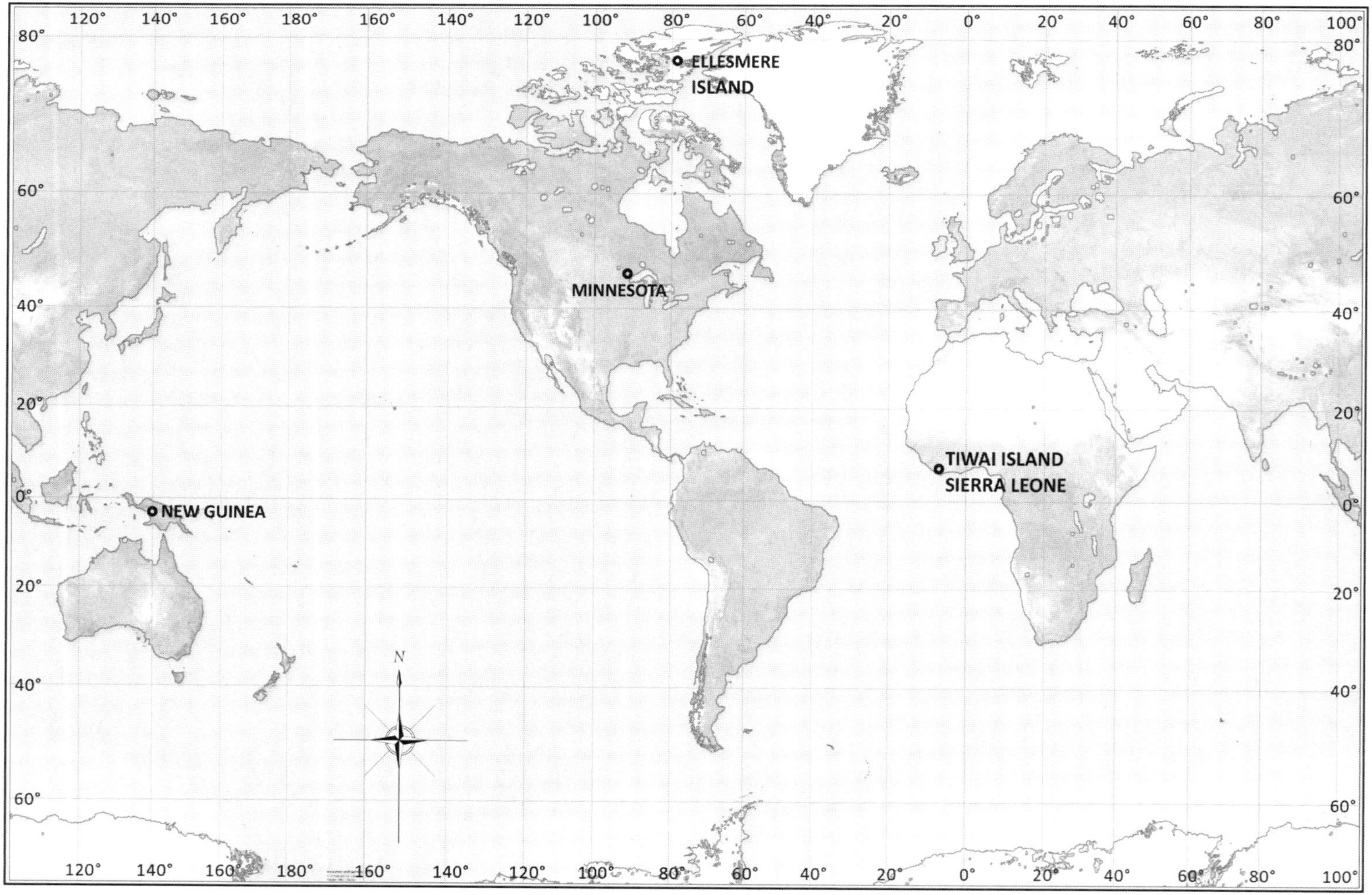
ELLESMERE ISLAND
MINNESOTA
TIWAI ISLAND SIERRA LEONE
NEW GUINEA
N

Gordon Buchanan

o

Low-lying and sparsely wooded, Calve Island stretches for about a mile across the mouth of Tobermory Bay on the Isle of Mull. Residents of the harbour town of Tobermory can gaze eastwards, across Calve Island to the mainland of western Scotland, only a couple of miles distant, while boats battling rough weather in the Sound of Mull can head west around Calve Island to find a quiet anchorage on the other side.

For one young lad growing up in Tobermory, Calve Island represented the pinnacle of his ambitions. Since the age of seven, when his family moved to Mull, Gordon Buchanan had spent pretty much every waking hour outdoors, either alone or with friends. He'd got to know the local woods and the wild coastline, collecting frogs and newts and slow worms, and when he wasn't taking part in an adventure he was dreaming of one.

Gordon's imagination was captured by the vessels that called into Tobermory harbour: he wanted to know where they were going, and felt a keen curiosity about far-flung places as yet unseen. That was for the future; for a lad still in primary school, the cravings of wanderlust were focused on Calve Island, and even that seemed unattainable. But luckily, one of Gordon's pals had a solution.

Norrie MacDonald's dad had taken Gordon and Norrie out a couple of times in his rowing boat, and before long the two lads had got into the habit of pottering around the harbour on their own. That was safe enough; but getting across the half-mile of water that lay between them and Calve Island felt like the equivalent of the lunar landing. It would be risky, but that only added to the allure. For company - or perhaps for comfort - they took Norrie's uncle's dog with them.

Brimming with all the exuberance of a long-anticipated lark, the two seafarers struck out across the bay. As they left the harbour, something of a more serious note skimmed briefly through Gordon's mind as he gazed down into the darkness of the deepwater channel they were crossing, and it occurred to him that no one actually knew where they were. But with the same kind of luck as Apollo 11's landing craft drifting down into the Sea of Tranquility, they glided safely across the intervening channel, poled through the tangled masses of seaweed that fringed the shore of Calve Island, and scrambled out onto the rocks.

From this new perspective, Gordon stood and looked back at Tobermory, with its hotels and shops and brightly painted houses. Over there, everyone was getting on with their daily lives. And here he was, in a world apart from all of that, somewhere wild and rarely visited, with just a pal and a dog for company. All around him were nesting seabirds and wildfowl, while beneath his feet were the yellow catkins of bog myrtle. A whole island of unfamiliar territory, ripe for exploration. It was an intoxicating feeling.

Naturally, that first voyage was followed by another and another. The youthful pioneers combed the island's shores and picked up all kinds of curiosities: shells, driftwood, shards of sea-glass, cuttlefish bones, and even a small metal canister containing a message from ocean scientists, which set their imaginations on fire. At a deeper level, Gordon found peace, fulfilment, and a strong sense of belonging. Long before he possessed a camera or a pair of binoculars, these clandestine escapades helped shape the course of his life.

'The outdoors was my happy place back then, and it's still my happy place now.'

'The Isle of Mull,' says Gordon, 'is an extraordinary place for kids to grow up and explore. The outdoors was my happy place back then, and it's still my happy place now.' [1] By contrast, however, school felt like a form of captivity. 'I spent most of my time daydreaming,' he remembers, 'looking out of the window, and thinking about what I was going to do once I actually got out of school.' [1]

Perhaps he'd be a pilot, gazing down on herds of African elephants from a small plane? Gordon's daydream, inspired by a TV programme, collided with the suggestions served up by a school careers adviser. Gordon didn't want to be a children's nanny and he didn't want to be a zookeeper, much as he loved wildlife. Instead, when he left school he got a job at a pony trekking centre, and was relieved to find something that he was naturally good at. 'Working with horses,' he says, 'played a huge part in giving me confidence. To get the best out of horses and work with them, you have to look at the world through their eyes. And that's exactly what you have to do with wildlife.' [1]

For a time, Gordon was convinced he would work with horses for the rest of his life. But when he was 17, he was offered an opportunity that chimed so perfectly with his dreams that it was scarcely believable. Nick Gordon, a wildlife cameraman living on Mull, was in need of an assistant for his next assignment. The location was an island, sparsely inhabited and hard to access, but distinctly hotter and more exotic than Calve Island in Tobermory Bay.

Washed on all sides by the powerful Moa River, Tiwai Island lies some 37 miles from the coast of Sierra Leone. Its habitat constitutes part of the Upper Guinea Rainforest, which once stretched in a green swathe across West Africa from Ghana to Guinea. Today, nearly three-quarters of this rainforest has been claimed by logging and mining, but Tiwai Island was spared these ravages because the local Mende people regard it as a sacred place. Now a haven for rare and endangered species, Tiwai Island contains one of the Earth's highest and most diverse concentrations of primates, and it was on these residents that the film-makers intended to focus.

For a young man who'd never left Britain, the shock of arriving in Sierra Leone was intense. In the blistering heat, as he struggled to load heavy crates of gear onto a leaking ferry, Gordon silently derided himself for his stupidity. He had leapt at Nick's offer, knowing little, in the pre-Google era, of the contrasts that he was about to encounter. Within hours of landing in Freetown, he was appalled at the evidence of poverty and disease that he saw all around him. 'I was desperately homesick,' he recalls. 'But I stuck with it. I knew I couldn't jump ship and leave my boss without an assistant, so I just kind of got on with it.' [1]

The film-makers' primary goal was to record chimpanzees using tools, a practice not yet documented in the west of Africa, although it had been filmed in the east. Gordon built scaffolding towers, put up hides, installed camera equipment, and got on with his everyday chores of cooking and housekeeping within the camp. From a nearby riverbank, the sound of stones being bashed onto a wooden surface rang out with promising regularity, and it seemed as if the reels of groundbreaking footage were virtually in the bag. But for six months, Tiwai Island's troop of nut-cracking chimps cleverly evaded the lenses.

The situation was already desperate when Gordon had an encounter that stopped him dead in his tracks: a big male chimp was standing on the path right in front of him. The animal met his gaze, then casually plucked a stick from a tree, stripped it of leaves, and poked it into a termite mound before withdrawing it and licking off the insects. Gordon, unequipped with a camera, could scarcely believe his eyes: these chimps were evidently multi-skilled. But a first-hand account with no evidence wasn't going impress TV producers, and many more weeks of fruitless waiting in a hide had to be endured before the animals were filmed wielding stones to crack nuts, using a tree root as an improvised anvil.

The documentary about the chimps of Tiwai Island extracted monumental levels of effort and commitment from its creators; despite the long months of homesickness and discomfort, and the shock of having to flee the country when it erupted into civil war, Gordon realised that he had done his job well. His next big question was whether he wanted to continue doing it, and ultimately become a cameraman himself. He spent two years helping to make a pioneering wildlife documentary in Venezuela and Brazil, and afterwards he knew the answer. He invested in his own camera gear and started looking for work.

Several years and a clutch of wildlife documentaries later, Gordon was on the lookout for leopards in Sri Lanka. These notoriously elusive predators were only putting in brief appearances, so the producer suggested that Gordon supplement the footage by making a video diary of his attempts to film them. 'It wasn't part of the plan,' explains Gordon. 'Other than Sir David Attenborough's documentaries, I wasn't a fan of presenter-led wildlife programmes. I felt that I was going to be ruining what could be a really good film by having a presenter pop up, but there was no way around it, really - it was either that or be left with half a film.'[1]

Because this wasn't something that he was desperately keen to do, Gordon deliberately went about it in his own way. His informal, conversational approach gave audiences an insight into the extraordinary rigours faced by a wildlife cameraman in order to get the shots he wants; the long hours of discomfort, the frustration, the disappointments and the sudden joy, all shared in an unobtrusive whisper, added a very relatable human narrative. The documentary was well received, and other commissions followed, including the BBC's *Lost Land...* series, which transported a team of scientists and film-makers into remote and little-known regions and recorded their discoveries.

Looking back at the making of *Lost Land of the Volcano* in New Guinea, Gordon is still astonished at the richness of the flora and fauna that the team encountered. Initially, he'd been apprehensive: as the helicopter prepared to land, he'd gazed down at the vast panorama of forested mountains and twisting brown rivers and knew that this was unlike anywhere he'd been before. By this stage in his career he'd travelled widely, but the terrain around the extinct volcano of Mount Bosavi looked daunting and impregnable. How was he going to locate the wildlife, let alone film it?

But within hours of the team's arrival, a Doria's tree kangaroo, a species found only in Papua New Guinea, ambled casually past the camp, and Gordon was soon filming it feeding in a tree from just seven metres away. He was enchanted by the behaviour of a pair of pygmy parrots, smaller in size than some of the rainforest insects, which he filmed at their nest in a termite mound; and he watched in awe as king birds-of-paradise performed their exquisite mating displays in the trees.

Among the 40 or so species and subspecies discovered during the making of *Lost Land of the Volcano*, one of the undisputed highlights was Gordon's encounter with the Bosavi woolly rat. This was an entirely new species to science, weighing 1.5 kilos and measuring 32 inches from nose to tail. The animal was filmed at night amid dense, rain-drenched undergrowth, and it continued foraging for food with no apparent concern for the onlookers crouching around it with their head torches.

For a wildlife film-maker who was used to concealing himself in a hide and waiting for something to happen, the extraordinary approachability of the creatures around Mount Bosavi was an almost surreal experience. Gordon explains: 'Because [the volcano] was such a hard to reach place, the species there were oblivious to human beings. You realise that fear is something that is not just learned and experienced, it's something that's built up over thousands and thousands of years.' [1]

While a fear of humans is ingrained in most animal species, a fear of certain animals is also inherent in us. As a presenter, Gordon frequently confronts his own fears in order to find out more about the animals he's filming and to challenge preconceptions about their nature. 'To better understand any animal,' he writes, 'you have to inhabit its world.' [2]

> ***'To better understand any animal, you have to inhabit its world.'***

In the forests of Minnesota, while making the BBC documentary *The Bear Family & Me*, he followed two different female black bears who were suckling their cubs. In time, the bears grew to tolerate his respectful company, and his own nervousness subsided. Speaking about one of the female bears, with whom he'd developed a special trust over the course of a year, he told viewers: 'I was terrified of this bear and now all that fear has gone. I'll never think about these animals in the same way again... I really hope I've done something to persuade other people that bears aren't the monsters we assume them to be.'[3]

Living for long periods in close proximity with a wild animal often leads to a finely-tuned awareness of that animal's attitude and behaviour; this can develop into a kind of sixth sense, forewarning film-makers of an imminent event, and allowing them to respond with great intuition. At no time did Gordon draw on this sense more deeply than when he was in Ellesmere Island, filming a pack of Arctic wolves.

Within 24 hours of pitching camp on the tundra, he was staring into the brilliant eyes of a wolf that had ventured just an arm's length from his lens. She had likely never seen a human before, and she had come to investigate the newcomer. On his part, Gordon had to quickly set aside the surprise of a wolf seeking him out, and judge for himself what to do next. In that moment, as he ignored his pounding heart and met her gaze with a semblance of calm, he could see her keen intelligence at work. In fact, each was silently reading the other with uncanny similarity.

In the days that followed, other wolves came and went, wary but inquisitive. It was an unnerving situation, but Gordon found himself smiling all the same. He watched the wind ruffling their fur and studied the size of their claws. It was impossible to forget that, if they chose to attack, these animals would be unfazed by the thin strand of electric fence that he'd strung around his tent. But something deeper was going on, and Gordon became more confident in their presence; he was seeking, and slowly finding, that vital balance between vulnerability and intrusion.

And meanwhile, how would he deal with a bold young wolf who gnawed at the camera traps, brazened his way past the fence and made off with a camp stool? Gordon realised that if he didn't act quickly, his tent could be ripped apart and his gear would be fast disappearing over the horizon. So he followed the retreating looter across the tundra, making what he hoped was a series of convincing howls which he changed into submissive moans when the wolf turned to face him.

At close range, Gordon could see the thick muscles of the wolf's neck and back; he, the film-maker, was now the subject, and he was being circled by a creature that was both beautiful and wild. Exhilaration mingled with more rational thoughts of alarm. Instinct had got him to this point: after all, the wolf showed no sign of aggression. Instead, he just seemed insatiably curious. Gordon spoke to him calmly, assuringly, keeping eye contact. After a while, as if throwing in his cards with both Gordon and the disputed stool, the wolf turned and stalked away.

By the end of his first stay on Ellesmere Island, Gordon had found and filmed a den with three pups. Edging closer, he'd filmed their mother, whom he called Luna, feeding them; he had watched the adult wolves go out hunting, and he'd witnessed their return. He had even, he felt, gained their trust to the extent that on one occasion the entire pack, including the cubs' mother, went off to hunt and left the cubs behind, in the knowledge that Gordon was nearby. They'd accepted his presence, and in turn, he had adapted his own behaviour: instead of watching them constantly, he carried on with his daily activities as if living with a pack of wolves was something that he did all the time.

Returning in the autumn of that year, Gordon was relieved to find the pack intact and the cubs looking strong and independent. He could tell that they recognised him, but now they were all on the move, hunting more widely and fattening themselves up for the coming winter. He followed them by quad bike, and when he struggled to navigate ravines he would always find them waiting for him to catch up. When the time came for their ways to part, he knew an almost overwhelming impulse to slip the leash that was tethering him to his own world and stay with the wolves as they headed across the tundra.

For a film-maker, the establishment of any kind of bond with animal subjects is never guaranteed, nor do events always follow the best possible course. Gordon has been chased - on different occasions - by bears, elephants and tigers, and says that at such times he's discovered how fast he can run. His most serious injury, however, has been a bite to his finger, delivered by a capercaillie in Scotland's Caledonian forest. In a hide, he never forgets that he is pretty easy pickings for any predator, but in defence of wild animals he argues that the combined annual tally of human fatalities caused by sharks, crocodiles, hippos, elephants, lions and bears is but a small fraction of annual deaths by homicide.

Gordon is unsentimental about filming predators killing prey, and has only rarely stepped in to save a vulnerable animal from preventable death. On witnessing a black bear cub abandoned by her mother, he and his team made the decision to leave out food for her, so that she didn't starve to death before the spring. 'Black bears,' he explains, 'are hunted in North America, and many more are hit by cars - we're having a massive impact on the longevity of these individual animals. Morally, if we could help this one little bear survive its challenges, it felt like the right thing to do.' [1]

Inevitably, Gordon has seen dramatic changes in the environments he's visited during his career. He says: 'To distil it into one problem, it's overpopulation by humans. In the Amazon, I've seen a huge amount of rainforest lost since I first went there in the early 1990s. It's saddening to fly over the edge of the Amazon and look out. Thirty years ago, it would have been unbroken forest for hundreds and hundreds of miles, but now it's just cattle ranches and lonely individual trees. And in Kenya, there's now a sprawl of humanity in a country that was at one time incredibly wild and rugged… huge areas have been fenced off, interrupting the natural migration patterns of lots of species, and settlements that were maybe a few shacks fifteen years ago are now bustling towns with all the problems that come with people occupying these previously wild places.' [1]

He does, however, see positive signs. 'If you'd asked me thirty years ago, when I first filmed tigers in the wild, whether I thought they'd still exist in the wild in 2024, I'd have said there was little chance. But happily, there are more tigers alive today than there were then. Tiger conservation is one of those success stories, although there's still a long way to go. And closer to home, when I drive around Scotland and see the results of rewilding programmes, where places are once again covered with trees, it really sings to my heart because it proves that with a few small steps and some committed individuals you can make a real difference.' [1]

He adds: 'There's a lot of hope to be taken, just from the ways in which we're viewing nature differently and valuing it properly for the first time. Back when I started film-making, deforestation, loss of habitat and climate change weren't even terms that were recognised.' [1]

In 2024, Gordon embarked on a journey across the Gobi Desert, braving freezing night-time conditions and deep snowfall on high mountain passes. Keeping a watchful eye out for wildlife, he juggled with two single-minded Bactrian camels, who didn't hesitate to make a bid for freedom when he forgot to hobble them, but whom he eventually grew to love. During his trek he met nomads who shared a glimpse of their traditional way of life, giving him a fresh perspective on the concepts of freedom and happiness. 'The closer you are to nature,' he said, 'I think the better you can understand yourself, because we surround ourselves with stuff that we don't need… There's a lot to envy about the contentment that people here seem to have.' [4]

> *'There's a lot of hope to be taken, just from the ways in which we're viewing nature differently.'*

Gordon has often spoken about the pleasure that comes from

sharing his work with a wide audience, especially when it helps to raise our awareness and appreciation of the natural world. That can only be of long-lasting benefit, because our instinct is always to protect the things that we care about. But at a more personal level, he finds fulfilment simply by watching wildlife in all its colour and complexity, and he's fascinated by the details as well as the drama. In a hide, there are no distractions; every moment receives his full attention. In spirit, he's the same rebellious lad who rowed off to Calve Island just because he wanted to see what was there.

[1] Interview with Jo Woolf (December 2024)

[2] Gordon Buchanan, *In the Hide* (2025)

[3] *The Bear Family & Me*, documentary directed by Jonny Young (2011)

[4] *Gordon Buchanan's Epic Camel Adventure*, documentary directed by Jack Warrender (2024)

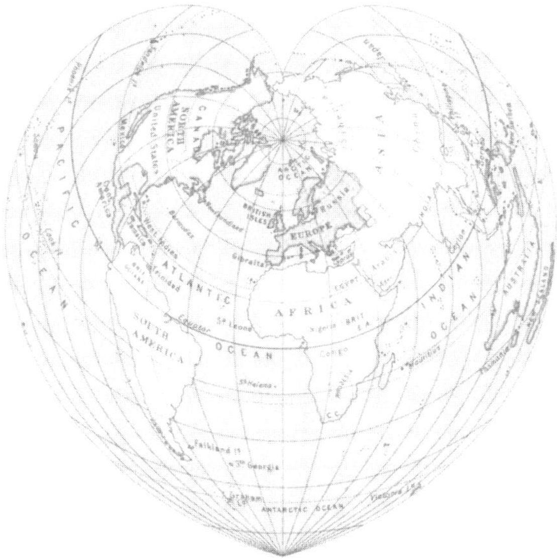
NORTH AMERICA
United States
PACIFIC OCEAN
BRITISH ISLES
EUROPE
West Indies
Gibraltar
ATLANTIC OCEAN
AFRICA
SOUTH AMERICA
Equator
INDIAN OCEAN
AUSTRALIA
Egypt
Nigeria
St Helena
Mauritius
Falkland Is
Sth Georgia
ANTARCTIC OCEAN

Mike Robinson

o

For over 25 years, Mike Robinson has been one of Scotland's leading figures in the field of climate change. It's a long time for anyone to be addressing an issue that causes widespread feelings of grief and powerlessness, but Mike is in it for the long haul. 'Grief,' he says, 'is understandable but it's not helpful - it's tiring and disengaging.' [1] He agrees with the Norwegian philosopher Arne Næss, who believed that the only way to deal with one's sorrow at the state of the world was to do something about it. 'I'm not prepared to be disempowered,' he says. 'Just give me something I can do that will be helpful.' [1]

It was earlier in his life, while working in a job that he didn't enjoy, that Mike saw a poster inviting applicants for Operation Raleigh, a worldwide volunteering programme founded by Colonel John Blashford-Snell. Having survived the infamous selection process, Mike selected Borneo as a destination because he knew it would take him well outside his comfort zone. He remembers: 'I was afraid of quite a lot of things at that age. Heights, creepy-crawlies… and there are plenty of mountains and creepy-crawlies in Borneo. I thought the point of it was to challenge yourself, so I set about it with that frame of mind.' [1]

Mike was already aware of environmental issues: 'Respecting nature,' he says, 'just felt like common sense.' [1] As a child, he was particularly moved by reading *Last Chance to See*, in which author Douglas Adams and zoologist Mark Carwardine travelled the world in search of the world's most endangered animals. In Borneo, he was astonished to enter an area described as a national park, only to discover that it had a perimeter of secondary jungle as a facade, concealing an area that was bare of trees. Logging activities were going on everywhere. 'It was,' he says, 'very easy to see environmental destruction. The rivers ran brown because of the amount of deforestation… the government was permanently having to de-silt the river mouths.' [1]

'Respecting nature just felt like common sense.'

But Mike was also struck by the way in which the people of the rainforest lived their lives. The entire community shared one longhouse, with a street and a string of homes under one roof. Their lifestyle was basic but comfortable, and they depended on their knowledge of the forest in order to survive. On finding out that Mike was a vegetarian, one of the guides, a Dayak tribesman called Felix, responded by taking him out into the forest in order to find things he could eat.

Mike recalls: 'Felix had a use for every single plant in the jungle... and when I thought I'd found one that he had no use for, which was a bush with incredibly hairy bark, it turned out that they snapped the twigs into pieces and used them as a children's game! I was absolutely bowled over.' [1] The villagers were careful not to over-harvest: if a species became hard to find, a decision was made to stop hunting or picking it until it had recovered. 'It wasn't a perfect symbiosis,' says Mike, 'but it was quite sustainable and fairly egalitarian.' [1]

When an injury forced a temporary evacuation to a camp in the north, Mike got to see a different side of the country. Villagers lived in individual houses, some of which lacked a wall or a roof; meanwhile the elders were employed by the logging company, meaning that they possessed a truck or a quad bike. Mike recalls: 'It was difficult not to be moved by the inequity of it. This reaffirmed to me how awful logging was, socially as well as environmentally. In every sense it was ripping the heart out of the place.' [1]

Having set out with the intention of facing one set of challenges, Mike arrived home with another. Many injustices were evident in Borneo, and he'd made a promise to Felix that he would try and do something about them. But how? He started off by writing to every charitable organisation that dealt with rainforests, asking them what action they were taking in Borneo. Just one responded: Survival International. But from their reply, Mike quickly realised that the destruction of rainforests wasn't limited to Borneo: industrial logging was happening in almost every rainforest on the planet.

Mike's time with Operation Raleigh had given him confidence and a clearer sense of purpose. He began to involve himself with Survival International, inviting tribespeople to Edinburgh and hosting political and media events. Ultimately, he became a board member of Survival International; in the more immediate future, his professional career led him into the field of fundraising for conservation projects. It was while working for RSPB that one individual in particular expressed a violent dislike to his approach.

'He was a bit bonkers,' recalls Mike. 'I knew he was there, but I didn't expect him to go for my kneecaps.' [1] A male capercaillie, in all his fan-tailed, ferocious splendour, launched his surprise attack in the Abernethy Forest. Capercaillie are renowned for their impressive territorial displays, and this one was incensed at the sight of some humans out for a stroll in his ancestral kingdom.

'At first,' explains Mike, 'it just seemed like a crazy Scottish bird.' [1] But Scotland's capercaillie population is declining, and it was suggested that a dwindling gene pool could lie at the root of this aggressive behaviour. At the time, it was known that there were six regions of Scotland's pine forest that held capercaillie, but because the forest itself was so fragmented, these regions were no longer connected. To help address the problem, Mike began talking to landowners with the aim of creating forested 'stepping stones' that would reunite the scattered population from Spey Bay to Kinlochleven. Meanwhile his attacker*, perhaps heartened by the sight of fleeing hikers, was forcibly removed by police after menacing shoppers in the streets of Grantown on Spey.

From taking on successive projects, Mike quickly learned that addressing one problem in isolation was never a perfect solution: 'There was always something else you needed to do... in the big scheme of things, it was never big enough.' [1] And during his time at RSPB, one issue in particular was still hovering on the fringes of most people's awareness. He says: 'I realised that nobody was talking about climate change. But somebody, somewhere, had to.' [1]

Mike's story of how he came to be offering solar panels to RSPB members goes something like this: he was flicking through the charity's gift catalogue, and was appalled to see a Beatrix Potter pyjama case, priced at £45. People obviously needed something more useful to spend their money on, and an investment in solar energy seemed like an obvious alternative. Underlying this scenario was Mike's deepening sense of concern at reports of rising global temperatures, fuelled by the emissions of a burgeoning human population.

The major hurdles were practical ones: translating the only available literature about solar panels from German into English; facilitating UK-wide fitting, which involved the cooperation of plumbers, electricians and electricity supply companies; and negotiating the availability of loans to finance it. Mike remembers: 'There was problem after problem, which is why I'm always drawn to Shackleton's famous quote: "Obstacles are just things to overcome, after all."' [1]

Initially, RSPB members were unconvinced; some demanded to know what solar energy had to do with the protection of birds. Mike wrote to them all politely and explained exactly what it had to do with birds, and with the rest of life on Earth. 'Invariably,' he says, 'the penny always dropped. It was just that they hadn't understood the connection. For me, this was a really valuable lesson: if you've thought something through and have good rational reasons for doing it, the first thing to do when somebody challenges you is to explain. The project resulted in around 1,200 people adopting solar energy at a time when it was desperately needed, and it also got my charity to start talking about climate.' [1]

As the signs of climate change became more evident, public concern increased. But, says Mike, 'despite real grassroots anxiety there didn't appear to be an appropriate public response.' [2] Then, in 2006, a climate bill was considered by the UK Parliament, influenced by a network of major environmental and humanitarian organisations. The network was called Stop Climate Chaos (SCC), and soon there was an appetite for a similar coalition in Scotland.

Stop Climate Chaos Scotland (SCCS) was established by a group of environmental and humanitarian NGOs, and Mike was invited to act as its chair. By this time he'd read hundreds of scientific papers and scores of books and reports on climate change. 'I viewed it as a pain barrier,' he says. If you could take in the enormity of the issue and still come out the other side with the energy for a fight, then you had made it through. You also needed enough belief that there was still time to take action… I didn't want to preach, depress or disenfranchise people, but rather encourage and motivate them to act.' [2]

In his next professional role, as Development Director at the Royal Botanic Gardens in Edinburgh, Mike helped organise a programme of talks for SCCS by leading experts, and the fledgling coalition began rallying volunteers and sharing information, which sparked media interest. He says: 'In that first year there was a lot of listening, enthusing, galvanising… we would speak for or with any group, congregation, community or individual who asked or wanted to help. The response was overwhelmingly positive.' [2]

In the years that followed, Mike went on to build SCCS into the largest coalition ever formed, bringing together 75 organisations with over 1.5 million members. SCCS provided a strong civil society voice that fed into the development of Scotland's first ever law on climate change: the Climate Change (Scotland) Act 2009. The first of the Act's 18 amendments set out the intention to achieve a 42% reduction in greenhouse gas emissions by 2020. This, says Mike, 'showed the need to take action early… In 2020, Scotland delivered a reduction of between 52% and 56% on 1990 levels. It had been viewed as out of reach and impossible, and then in 2020 we exceeded it.' [1]

The Act meant that Scotland was on the way to being recognised as a global leader in addressing climate change. SCCS also lobbied the Scottish government to agree to set up the world's first Climate Justice Fund, the first Loss & Damage commitments, and to create several other significant pieces of legislation. Fresh targets, including a 75% reduction in emissions by 2030, were set down by the Scottish Parliament in 2019, negotiated by SCCS and propelled largely by the impetus of worldwide school strikes and protests by young people who demanded immediate, coordinated political action on climate and the environment.

Then, in the spring of 2024, it was announced that the Scottish government was dropping its targets for 2030. In response, Mike wrote: 'We need to get back on track. Every single sector of society needs to step up and get back to the level of commitment, energy and focus that we had in 2019... This moment should be a wake-up call for every one of us... Do we actually care about the future or don't we?' [3]

'Do we actually care about the future or don't we?'

With hindsight, he now adds: 'Unfortunately, [the decision] has been portrayed as a failure of ambition... That is not true. It is a failure of action. There's nothing wrong with the ambition. But it was inevitably going to fail within a couple of years because government and industry needed to start doing things differently, and quickly, to have the necessary impact.' [1] In real terms, although the 75% target won't be reached, there is a chance that 70% is achievable. Mike is philosophical: 'It's frustrating, but it has also got to be kept in perspective. We are still a hell of a lot further forward than we were 25 years ago.' [1] And meanwhile he sees a need to remind politicians that most people want to see real change.

Mike has been Chief Executive of the Royal Scottish Geographical Society (RSGS) since 2008. The Society has a history stretching back to 1884, and is a repository of some of the most significant and exciting stories in the history of geographical exploration and discovery. RSGS has always been a staunch advocate of geography, both as an academic subject and in its application to the understanding and interpretation of current issues. 'There is,' says Mike, 'an ever-present weight of responsibility in running an organisation like RSGS with such heritage. How do we continue that legacy without slipping into becoming a historical society? Our job is, after all, to remain relevant and remind people of the importance of geography in the everyday.' [4]

RSGS members don't need convincing about the importance of geography, as most of them are formidably well-informed and well-travelled. Mike likes to recall an RSGS talk in Perth, in which a speaker asked the audience how many of them had ridden on a camel; almost two-thirds of them raised their hand. On another occasion, when Mike was hosting a public event for a Yanomami tribal elder from Brazil, one of the audience politely asked if they could speak to the guest in Yanomami.

In support of the ongoing role of geography in the modern era, Mike writes that it is 'a science with an understanding of the realism of living daily life... almost every major issue of our day has a strong geographical component and can be solved better through geography. I firmly believe it has a huge role in helping guide the future.' [4] RSGS has advisory roles on Arctic policy, aviation, citizen engagement and rail travel, and chairs external forums in fields such as agriculture and city development.

Through a number of initiatives established and supported by RSGS, geography is helping to shape our response to climate change on a national and international scale. Climate Solutions is an educational project designed to improve the 'climate literacy' of managers in the public, private, academic and voluntary sectors. In addition, RSGS hosts regular gatherings of the international geographical community, promoting greater cooperation in understanding and addressing the climate and biodiversity crisis. One such meeting in June 2021, hosted in partnership with the Royal Geographical Society (with IBG) and the International Geographical Union, resulted in a joint declaration of more than 80 societies worldwide, each pledging to redouble their efforts in addressing the climate emergency.

In many ways, both the study of geography and the debate about climate change demand an appreciation of shifts that can happen over decades, centuries and even millennia. No editor who puts a date of 'Spring 02024' on a quarterly magazine can be accused of short-term thinking. It's good to imagine the Earthlings of 10,000 years hence rifling through their back issues of *The Geographer* - impossible to imagine that some won't have kept hard copies - to read articles on the theme of long-mindedness and the geography of time.

In that same issue, Mike writes that 'a better understanding of the geography and philosophy of time is more vital than ever. By most measures we are failing to hand over the world in a better state than we inherited it.' [5] He points to Professor William MacAskill's book, *What We Owe The Future*, which encourages us to imagine that we're at the beginning of human life on Earth, and not, as is often believed, at the end. Mike asks: 'How would that affect our sense of responsibility to the future, and the billions of people not alive yet, who rely on some of the decisions we make today?'[5]

And alongside those decisions, in the face of climate change and other threats to the natural environment, it is also essential that we are willing to adapt. This response, he writes, 'is not just a necessity but an opportunity, and if we collectively get this right, we can at the same time improve the lives of people and communities around the globe.'[6] With this cooperation and flexibility, Mike believes that the decade of the 2020s has the potential to be productive, innovative, creative and purposeful: 'It will be a decade of altruism and empathy, responsibility and compassion... It will be a decade of burgeoning wildlife populations... a decade of improving food, air and water quality... of forests that are regenerating, not disappearing. Of seas teeming with life.'[7]

Mike is a keen hill walker and climber; he has climbed in Peru, Tanzania and the Himalayas, and has completed the round of Scotland's Munros. He says: 'Climbing, for me, requires extreme focus, a certain familiarity with failure and mortality, which in turn breeds humility and perspective.'[1] He emphasises the importance of our bond with the landscape and the natural world: 'Why do we relax when we see the sea or take time to observe the sky? Why do we all sense that mountains, forests, rivers and lochs contain such beauty? Why do we feel joy at the splash of an otter or the flash of a kingfisher? It feels spiritual and primal, like it's a basic need and not some luxury.'[1]

In 2017 Mike swam across the Gulf of Corryvreckan, the sea channel between the islands of Jura and Scarba which contains the world's third largest whirlpool. According to legend, the Corryvreckan is named after a Norse prince, Breakan, who accepted a challenge to anchor his ship in the whirlpool for three stormy nights, but both he and his ship went down on the third night when the rope snapped. By contrast, Mike's immersion in the Corryvreckan was timed to coincide with calm conditions, and it took about half an hour to swim the two kilometres from Jura to Scarba. 'Sometimes,' he writes, 'I just have to go off and do something fun purely for my own sanity.'[2]

'I get to see a lot of the good that's going on in the world.'

Finding solutions to problems; salvaging positivity out of a crisis; above all, promoting action over inaction. One of the benefits of Mike's ethos is that it continues to bring him into contact with like-minded people. He says: 'I get to see a lot of the good that's going on in the world, and I get to meet a lot of the people who are actually out there doing it.'[1]

Citing the comedian Billy Connolly, who once remarked that everywhere he goes in the world he sees people smiling and laughing, Mike believes that if you approach life in a particular way, that is precisely what you see reflected back. 'Just trying to be positive taps into groups of people doing really good things. I'd like to think I'm trying to do that in a magnanimous way, so that I can help other people know about it, or offer support. Instead of going round in circles, it becomes a different conversation - interesting, inspiring, progressive. How much more exciting is that?'[1]

*It is entirely possible that this is the same capercaillie who bit Gordon Buchanan. His offspring may still cherish a sense of pride that their ancestor tried to sabotage at least two successful careers.

[1] Interview with Jo Woolf (December 2024)

[2] Mike Robinson, *A View from a Chair* (unpublished memoir)

[3] Mike Robinson, 'A Retrograde Act', blog post for RSGS (19th April 2024)

[4] Mike Robinson, 'Community of Geography', *The Geographer* (Summer 2024)

[5] Mike Robinson, 'Why the Geography of Time Matters', *The Geographer* (Spring 02024)

[6] Mike Robinson, 'Adaptation is Common Sense', *The Geographer* (Winter 2024)

[7] Mike Robinson, 'A Decade of Hope', *The Geographer* (Autumn 2021)

About the Royal Scottish Geographical Society

Our mission: RSGS is a dynamic educational charity with far-reaching influence and a mission to inspire and inform. By thinking geographically, championing good practice, working with leaders and experts, and engaging with public audiences, we are precisely placed to inform policy, influence joined-up decision making, and inspire action, all to help address national and global challenges. We use our networks, knowledge and reputation to shape positive, practical policy solutions.

Our vision: RSGS promotes greater geographical understanding of people, places and the planet. In our increasingly interdependent world, and faced with complex environmental and social challenges, joined-up geographical thinking is needed as never before. Geography connects people and unites disciplines to help resolve many of our most pressing concerns, from climate change to food systems, from unsustainable transport to health inequalities, from urban design to biodiversity loss.

What we do: We produce a quarterly magazine, each exploring key geographical issues; we run an annual programme of around 100 public talks by inspiring people from the worlds of science, travel and adventure; we produce an academic journal; create educational resources; convene our network to help explain and solve complex problems; house a map and archive collection; advise government and industry; and work with an array of partner bodies nationally and internationally to inspire, inform and influence positive change. RSGS is also home to a number of creative projects including Climate Solutions - a scheme to improve the 'climate literacy' of managers in the public, private, academic and voluntary sectors.

RSGS is a membership charity. To find out more, and to become a member, visit: https://www.rsgs.org/

RSGS Connections

The RSGS has a long history of awarding medals and Honorary Fellowships (FRSGS). These recognise, in the broadest sense, a significant contribution to geography. The exact criteria for each award can be found on our website, together with a full list of recipients.

In addition, we have (so far!) hosted over 6,000 talks by leading figures in science and human endeavour, offering what amounts to a front-row seat on many of the most exciting and inspirational moments in the last 150 years of geographical exploration and discovery.

All of the people featured in this book have some connection to the RSGS, either directly or indirectly. Below is a brief summary of each, with additional notes of interest.

John Rae, FRS, FRGS is remembered in the name of Rae Strait, which lies between King William Island and the Boothia Peninsula in Nunavut, Canada. Rae was a member of RSGS from 1885, the year after the Society was founded. He died at his home in London in 1893, and was buried at St Magnus' Cathedral in Kirkwall. In 1895, when RSGS staged a special exhibition to commemorate the 50th anniversary of Sir John Franklin's lost expedition, Rae's widow, Kate, loaned a number of Franklin-related relics that Rae had obtained in the Arctic.

Carsten Egeberg Borchgrevink, FRSGS was made a Knight of the Royal Norwegian Order of St Olav by King Oscar II. Several features in Antarctica bear his name, including the Borchgrevink Coast and the Borchgrevink Glacier in Victoria Land. He received the RSGS Silver Medal in 1900, and was a guest lecturer in 1898 and 1900. He visited RSGS on other occasions too, signing his name in the Visitors' Book for the last time in 1930.

Sir Ernest Shackleton, CVO, OBE, FRGS, FRSGS received many official awards and decorations. His Polar Medal bears three clasps, corresponding to his first three Antarctic expeditions. His fourth expedition to the Antarctic, on board the *Quest*, was his last, and he died in South Georgia on 5th January 1922. Shackleton has a strong connection with RSGS: in January 1904, aged 29, he was appointed Secretary, and he stayed in the role until July 1905; he remained an RSGS Council member for several years thereafter. Shackleton was awarded the RSGS Livingstone Medal in 1909 and gave many talks for the Society between 1903 and 1920.

The Endurance22 Expedition: In March 2023, RSGS welcomed Dr John Shears, Nico Vincent and Natalie Hewit of the Endurance22 expedition and Donald Lamont, Chairman of the Falklands Maritime Heritage Trust, who together gave a special presentation in Perth Concert Hall about the discovery of the *Endurance*. More details were shared on 5th December 2024 by John Shears and Nico Vincent at the Festival of Shackleton in Dundee, hosted jointly by RSGS and The Dundee Heritage Trust.

Myrtle Simpson is a lifelong advocate for exploring the outdoors. She received the Polar Medal in 2017, in recognition of her Arctic expeditions. She is a former President of the Scottish Ski Club and her husband, Hugh, was a co-founder of the Scottish Arctic Club in 1970; on the Scottish Arctic Club's website, Myrtle is described as a 'founding gatecrasher.' Both Myrtle and Hugh were awarded the RSGS Mungo Park Medal in 1969. Myrtle has given many talks for RSGS, and most recently an interview at her home in the Highlands.

Hazel and Luke Robertson are motivational speakers and mindset coaches, as well as outdoor guides and enthusiasts. After Luke became the first Scot to ski solo, unassisted and unsupported to the South Pole, their joint expeditions in Alaska and the Arctic, and their completion of the

Marathon des Sables, Hazel and Luke were appointed RSGS Explorers in Residence. They have carried the RSGS flag to far-flung locations, and have given many talks for RSGS, both in person and online.

Colonel John Blashford-Snell, CBE, FRSGS has mounted over 100 expeditions worldwide and is the founder and supporter of numerous charities and community aid projects. John has received many awards and distinctions including the Segrave Trophy in 1974 and the Edward C Sweeney Medal of the Explorers Club in 1992. He was awarded the RSGS Livingstone Medal in 1974 and has given many public talks for the Society.

Benedict Allen, FRGS has written many books on his travels, and edited *The Faber Book of Exploration* (2002). He is a Trustee and Member of Council of the Royal Geographical Society. As a traveller, he has filmed a number of TV programmes, either solo or with camera crews. Benedict has been a guest speaker at RSGS on many occasions.

Sal Montgomery, FRSGS is a professional whitewater kayaker and a TV and film adventure specialist, often working as a researcher and safety consultant for production companies. Sal has been a guest speaker for RSGS on many occasions.

Anders and Anne Holch Povlsen, FRSGS were awarded the RSGS Geddes Environment Medal in 2023. This medal, named in honour of Sir Patrick Geddes, was also presented in 2015 to Dick Balharry, who featured in *The Great Horizon*; Thomas MacDonell is an Honorary Fellow of RSGS.

Ray Mears, FRSGS teaches bushcraft and survival skills through Woodlore, Britain's first school of wilderness bushcraft, which he founded in 1983. Ray was a guest of RSGS in October 2010, when he gave a presentation in Perth and was awarded the RSGS Mungo Park Medal.

Alexander Mitchell Kellas's pioneering ascents and groundbreaking research distinguish him as one of Scotland's greatest Himalayan mountaineers. His name featured prominently in a series of Mount Everest lectures given to RSGS by the climber George Mallory in 1922; Mallory paid warm tribute to Kellas, whose death in the Himalayas in 1921 was especially mourned in his home city of Aberdeen.

After the Scottish Women's Himalayan Expedition, **Monica Jackson** continued to be an active climber in Britain and abroad; she graduated in archaeology and anthropology from Cambridge University and completed a doctorate at Edinburgh University in 1976. **Dr Evelyn Camrass** (later McNicol) returned to her role as an obstetrician in Edinburgh, and was still an adventurous climber and skiier in her seventies. **Elizabeth ('Betty') Stark** trained as a speech therapist and moved to the US to begin a research career, where she became a leading expert on language development in young children. Betty gave a talk about the Scottish Women's Himalayan Expedition for RSGS in Edinburgh in December 1955.

Dougal Haston and Sir Chris Bonington visited RSGS in 1976 to receive the RSGS Livingstone Medal. This was awarded to the British Mount Everest Southwest Face Expedition of 1975, of which Bonington was the leader.

Doug Scott, CBE, FRSGS accumulated a dazzling record of first ascents and devoted his later life to aiding communities in the Himalayas, setting up Community Action Nepal in 1994. Doug received many awards and distinctions, including the Piolet d'Or Lifetime Achievement Award. As a member of the British Mount Everest Southwest Face Expedition, he shared in the RSGS Livingstone Medal of 1976. He gave many public talks for RSGS.

Polly Murray, FRSGS is an ambassador for outdoor adventure, and offers motivational presentations as a public speaker. She continues to ski and climb with her growing family. Polly has given a number of public talks for RSGS.

Leo Houlding, FRSGS is renowned for his expeditions on the world's toughest and most inaccessible walls of rock. As well as putting up many new routes and first ascents, Leo has worked with film-makers to retrace the last steps of George Mallory on Everest. Leo has been a guest speaker for RSGS on several occasions.

Gertrude Bell, CBE joined the British administration in Mesopotamia in 1917, and played a significant role in shaping the Kingdom of Iraq. Gertrude does not seem to have paid a personal visit to RSGS; however, Sir Percy Cox, Civil Commissioner (later High Commissioner) for Mesopotamia, to whom Bell served as Oriental Secretary between 1917 and 1923, gave a talk for RSGS in 1926.

Sir Marc Aurel Stein, KCIE, FRAS, FBA, FRSGS received many awards and distinctions and bequeathed a wealth of manuscripts and artefacts to institutions such as the British Library and the V&A. He became a British citizen in 1904. Stein was a Corresponding Member of RSGS between 1900 and 1901; he gave a series of lectures in 1910.

The Rt Hon Rory Stewart, OBE, FRSGS, FRSL is a former British diplomat and politician. He was a Member of Parliament for Penrith and the Border between 2010 and 2019, and from 2015 held a succession of ministerial posts. Rory has taught at Harvard University and is currently Professor of the Practice of Grand Strategy at Yale University's Jackson School of Global Affairs. He is an award-winning author, has presented several TV documentaries, and co-hosts the podcast *The Rest is Politics* with Alastair Campbell. Rory was awarded the RSGS Livingstone Medal in 2009 and has been a guest speaker on several occasions.

Mark Evans, MBE, FRGS, FRSGS has been leading expeditions for over 40 years, focusing on projects and capturing stories that leave meaningful legacies to both local and international communities and causes. In 2023 he was awarded The Medal of Honour by His Majesty Sultan Haitham bin Tariq Al Said, The Sultan of Oman, for his role in establishing Outward Bound in Oman. Mark has been a guest speaker at RSGS on many occasions.

Alice Morrison, FRSGS is an adventurer, TV presenter and travel writer, and she also produces a highly acclaimed podcast, *Alice in Wanderland*. Alice has been a guest speaker for RSGS on many occasions.

Sir Charles Wyville Thomson, FRSE, FRS, FLS, FGS, FZS died in 1882, two years before RSGS was founded, but with the Challenger Expedition's offices being based in Edinburgh, his legacy was still strongly felt. Sir John Murray served as President of RSGS between 1898 and 1904 and received the RSGS Livingstone Medal for his own contributions to oceanography. John Young Buchanan was awarded the RSGS Gold Medal (now the RSGS Scottish Geographical Medal) in 1911.

Sir Francis Chichester wrote several books about his journeys, and used his navigational expertise to publish a series of pocket guides and maps. He was awarded the RSGS Livingstone Medal before giving a talk in Edinburgh in 1967.

Doug Allan, OBE, FRPS, FRSGS has won eight Emmys, five BAFTAs and five Wildscreen Pandas, and has twice won the underwater category of Wildlife Photographer of the Year. He has received three honorary doctorates in recognition of his camerawork. Doug was awarded the RSGS Mungo Park Medal in 2018, when he gave a talk hosted by RSGS in Perth.

Lewis Pugh, OIG is the United Nations Patron of the Oceans and the recipient of numerous awards and honours, among them the Order of Ikhamanga (Gold Class), a South African civilian honour, in 2009. Through the Lewis Pugh Foundation, he continues to campaign for Marine Protected Areas in all of our oceans. Lewis was awarded the RSGS Mungo Park Medal and Honorary Fellowship in 2019, after giving a talk hosted by RSGS in Perth.

Bethany Hamilton is a professional surfer, a motivational speaker and the author of several books for adults and children on the themes of lifestyle, health and faith. Bethany was interviewed by Jo Woolf in 2020 for RSGS's magazine, *The Geographer*.

Elizabeth Maruma Mrema, FRSGS is Assistant Secretary-General of the United Nations and Deputy Executive Director of the United Nations Environment Programme. She has received a number of awards, including the Kew International Medal in 2022. Elizabeth was named one of the world's 100 Most Influential People by Time Magazine in 2023, and by the Financial Times as one of the 25 Most Influential Women of 2023. She was awarded the RSGS Geddes Environment Medal in 2024.

Bertrand Piccard, FRSGS has received many awards for his achievements in aviation, including the Harmon Trophy and the Hubbard Medal of the National Geographic Society. He is the Founder and Chairman of the Solar Impulse Foundation and continues to initiate and develop innovative solutions for a sustainable economy.

Likewise, **André Borschberg, FRSGS** has received many international recognitions, including The Explorers Club Medal and La Grande Médaille de l'AéroClub de France. He is the Co-founder and Executive Chairman of H55, which develops electric propulsion technology for aviation. In 2018, Bertrand and André gave a public talk hosted by RSGS in Glasgow, during which they were presented with the RSGS Mungo Park Medal.

Chris Packham, CBE has received many awards for his work in broadcasting and conservation, and serves as President, Vice-President or Patron of numerous wildlife and conservation organisations. He holds honorary doctorates from the University of Southampton and the Royal Holloway, University of London. Chris was a guest speaker for RSGS in December 2019.

Polly Higgins, FRSGS received a number of awards and distinctions for her work as a visionary environmental campaigner. She was awarded the RSGS Shackleton Medal in 2018, when she gave a presentation to a Masters class in Environmental Law at Strathclyde University.

Jojo Mehta, FRSGS is CEO and key spokesperson of Stop Ecocide International. She was interviewed by RSGS Chief Executive, Mike Robinson, for our online talks programme in 2022.

Gordon Buchanan, MBE, FRSGS has been making wildlife documentaries for over 30 years. In 2013 he received both an honorary doctorate from the University of Stirling and the Glenfiddich Spirit of Scotland Award in the environment category. Gordon was a guest speaker for RSGS in 2019.

Mike Robinson, OBE, FRCGS, FRSE was awarded the Royal Canadian Geographical Society's Captain Joseph-Elzéar Bernier Medal in 2017 and has received several other awards for work focused on climate and the environment, including the Glenfiddich Spirit of Scotland Award in 2009. In 2023 he received an honorary doctorate from the University of Stirling.

SELECTED BIBLIOGRAPHY

POLAR

John Rae
McGoogan, Ken, *Fatal Passage* (HarperCollins, 2001)
National Museums of Scotland, *No Ordinary Traveller: John Rae, Arctic Explorer 1813–1893* (1993)
Rae, John, *Narrative of an Expedition to the Shores of the Arctic Sea in 1846 and 1847* (T & W Boone, 1850)
The John Rae Society https://www.johnraesociety.com/

Carsten Borchgrevink
Bernacchi, Louis, *To the South Polar Regions* (Hurst and Blackett, 1901)
Borchgrevink, Carsten, *First on the Antarctic Continent* (George Newnes, 1901)

Sir Ernest Shackleton and The Endurance22 Expedition
Alexander, Caroline, *The Endurance: Shackleton's Legendary Antarctic Expedition* (Bloomsbury, 1998)
National Geographic Documentary Films, *Endurance*, directed by Jimmy Chin, Natalie Hewit, Elizabeth Chai Vasarhelyi (2024)
Huntford, Roland, *Shackleton* (Hodder & Stoughton, 1985)
Shackleton, Sir Ernest, *South* (William Heinemann, 1919)
Shears, John and Nico Vincent, *Endurance: The Discovery of Shackleton's Legendary Ship* (National Geographic, 2024)
The Endurance22 Expedition: https://endurance22.org/

Myrtle Simpson
Simpson, Myrtle, *Home is a Tent* (The Travel Book Club, 1964)
Simpson, Myrtle, *White Horizons* (Victor Gollancz Ltd, 1967)

Luke and Hazel Robertson
http://everyday-exploring.com/

FOREST & RIVER

Colonel John Blashford-Snell
Blashford-Snell, John, *From Utmost East to Utmost West* (Bradt Guides Ltd, 2022)
Blashford-Snell, John, *Something Lost Behind the Ranges* (HarperCollins, 1994)
The Scientific Exploration Society: https://www.ses-explore.org/

Benedict Allen
Benedict Allen, *Explorer* (Canongate, 2022)
Benedict Allen, *Mad White Giant* (Macmillan, 1985)
Benedict Allen, *The Proving Grounds* (HarperCollins, 1991)
https://www.benedictallen.com/

Sal Montgomery
Backshall, Steve, *Expedition* (BBC Books, 2019)

Anders Holch Povlsen
https://wildland.scot/

Ray Mears
Mears, Ray, *British Woodlands* (Ebury, 2023)
Mears, Ray, *My Outdoor Life* (Hodder & Stoughton, 2013)
Mears, Ray, *Northern Wilderness* (Hodder & Stoughton, 2010)
https://www.raymears.com/

MOUNTAIN

Alexander Kellas
Gillman, Peter and Leni, *The Wildest Dream: George Mallory, The Biography of an Everest Hero* (Headline Book Publishing, 2000)
Mitchell, Ian R, and George W Rodway, *Prelude to Everest: Alexander Kellas, Himalayan Mountaineer* (Luath Press, 2011)

The Scottish Women's Himalayan Expedition
Jackson, Monica, and Elizabeth Stark, *Tents in the Clouds* (Collins, 1957)

Doug Scott and Dougal Haston
Bonington, Chris, *Ascent* (Simon & Schuster, 2017)
Bonington, Chris, *Everest the Hard Way* (Hodder and Stoughton, 1976)
Connor, Jeff, *Dougal Haston: The Philosophy of Risk* (Canongate, 2002)
Haston, Dougal, *In High Places* (Cassell, 1972)
Scott, Doug, *Up and About: The Hard Road to Everest* (Vertebrate Publishing, 2015)

Leo Houlding
BritRock Films, *The Last Great Climb*, directed by Alastair Lee (2013)
Houlding, Leo, *Closer to the Edge* (Headline, 2022)
https://www.leohoulding.com/

Polly Murray
London Speaker Bureau: https://londonspeakerbureau.com/speaker-profile/polly-murray/

DESERT

Gertrude Bell
Bell, Gertrude Lowthian, *The Desert and the Sown* (William Heinemann, 1907)
Howell, Georgina, *Gertrude Bell: Queen of the Desert, Shaper of Nations* (Farrar, Straus and Giroux, 2006)

Sir Aurel Stein
Stein, M Aurel, *Ruins of Desert Cathay* (Macmillan, 1912)
Stein, M Aurel, *Sand-buried Ruins of Khotan* (Hurst and Blackett, 1904)
Walker, Annabel, *Aurel Stein: Pioneer of the Silk Road* (John Murray, 1995)

Rory Stewart
Stewart, Rory, *The Places in Between* (Picador, 2004)

https://www.rorystewart.co.uk/
https://www.turquoisemountain.org/

Mark Evans
Evans, Mark, *Crossing the Empty Quarter* (Gilgamesh Publishing, 2016)
Thomas, Bertram, *Arabia Felix* (Jonathan Cape, 1932)
https://exploringwithpurpose.uk/

Alice Morrison
Morrison, Alice, *Adventures in Morocco* (Simon & Schuster, 2019)
Morrison, Alice, *Walking With Nomads* (Simon & Schuster, 2022)
https://alicemorrison.co.uk/

OCEAN

Sir Charles Wyville Thomson
Campbell, Lord George, *Log-Letters from the Challenger* (Macmillan and Co, 1881)
Corfield, Richard, *The Silent Landscape: In the Wake of HMS Challenger 1872-1876* (John Murray, 2004)
Linklater, Eric, *The Voyage of the Challenger* (John Murray, 1972)
Thomson, C Wyville, *The Depths of the Sea* (Macmillan and Co, 1873)
Thomson, C Wyville, *The Voyage of the Challenger* (Macmillan and Co, 1877)

Sir Francis Chichester
Chichester, Francis, *Gipsy Moth Circles the World* (Hodder & Stoughton, 1967)
Chichester, Francis, *The Lonely Sea and the Sky* (Hodder & Stoughton, 1964)
Leslie, Anita, *Francis Chichester: A Biography* (Walker and Company, 1975)

Doug Allan
Allan, Doug, *Freeze Frame* (Tartan Dragon Ltd, 2012)
https://dougallan.com/

Lewis Pugh
Pugh, Lewis Gordon, *Achieving the Impossible* (Simon & Schuster, 2010)
https://www.lewispugh.com/
https://www.lewispughfoundation.org/

Bethany Hamilton
Hamilton, Bethany, *Soul Surfer* (2004)
https://bethanyhamilton.com

PALE BLUE DOT

Elizabeth Maruma Mrema
UN Convention on Biological Diversity: https://www.cbd.int/
Global Youth Biodiversity Network: https://www.gybn.org/

André Borschberg and Bertrand Piccard
Piccard, Bertrand, *A Trace in the Sky* (Editions Favre, 2008)

https://bertrandpiccard.com/
https://solarimpulse.com/
https://aroundtheworld.solarimpulse.com/
https://www.andreborschberg.ch/
Electric Propulsion Solutions by H55: https://h55.ch/en/

Chris Packham

Packham, Chris, *Fingers in the Sparkle Jar* (Ebury, 2016)
Packham, Chris and Andrew Cohen, *Earth* (William Collins, 2023)
Packham, Chris and Megan McCubbin, *Back to Nature* (Two Roads, 2020)
https://www.chrispackham.co.uk/

Polly Higgins and Jojo Mehta

Higgins, Polly, *Dare to be Great* (Flint, 2020)
Higgins, Polly, *Earth is Our Business* (Shepheard-Walwyn (Publishers) Ltd, 2012)
Higgins, Polly, *Eradicating Ecocide* (Shepheard-Walwyn (Publishers) Ltd, 2010)
Stop Ecocide International: https://www.stopecocide.earth/

Gordon Buchanan

Buchanan, Gordon, *In the Hide* (Witness Books, 2025)
https://gordon-buchanan.co.uk/

ACKNOWLEDGEMENTS

This book could not have been written - or even started - without the support, cooperation and generosity of so many people.

Firstly, my heartfelt thanks to my friends at RSGS HQ for their unwavering encouragement and support: Mike Robinson, Clare Hamilton, Katrina Strachan, Susan Watt, Holly McNair, Jo Donachie and Lindsey Gibb.

Likewise, warmest thanks to the RSGS Collections Team: Margaret Wilkes, Bruce Gittings, John Lewington, Pat Brown, Dorothy Degenhart, Blair White, Sam Chakraverty, Jean Smith, Jane Griffiths, Andrew Cook, Michael Cairns, and Colin Forsyth. I'd also like to acknowledge the fond memory of the late Kenny Maclean who was always so supportive of my work, and made many valuable suggestions over the years.

Thank you to RSGS's Chairman, Professor John Briggs and the Board of Trustees; and to RSGS members and supporters who have given me so much encouragement and helped to make this project a reality. A big thank you also to Jo Sharp, for writing the Foreword.

For their generous collaboration and input, in terms of either giving interviews, answering questions, supplying further information, or taking time to look over my text, my grateful thanks go to Doug Allan, Benedict Allen, Colonel John Blashford-Snell, Gordon Buchanan, Mark Evans, Bethany Hamilton, Leo Houlding, Ray Mears, Jojo Mehta, Sal Montgomery, Alice Morrison, Elizabeth Mrema, Polly Murray, Chris Packham, Bertrand Piccard, Anders Povlsen, Lewis Pugh, Hazel and Luke Robertson, Mike Robinson, John Shears, Myrtle Simpson and Rory Stewart.

For their assistance, I would also like to thank Simon Jackson; Alison Higham at the Ladies' Scottish Climbing Club; Paula Williams, Curator of Map, Mountaineering & Polar Collections, National Library of Scotland; Chris Fleet, Map Curator, National Library of Scotland; Professor Peter Nienow and Professor Noel Gourmelen at the University of Edinburgh; Joy Wheeler, Assistant Librarian (Photographs and Artworks) at the Royal Geographical Society (with IBG); Frances Daltrey, Chris Bonington Picture Library; Graham Robson, Special Collections & Archives, Newcastle University; Ali AlQahtani, Celicourt Communications; Sue Stockdale, explorer and host of Access to Inspiration podcasts; Mohamed Al Zadjali; Amur Al

Wahaibi; the team at Woodlore; Rob Monaghan, Stop Ecocide International; Dave Smith at the Scientific Exploration Society; Martina Tasev; Charlotte Corney; Charlotte Bellini; Katie Tolfree; Kristine Mønster; Tim Kirkwood, Wildland; Steven Liddle, Greenbeard; Clare Bean, Jo Sarsby Management; Sarah Rumboll, The Lewis Pugh Foundation; Mike Coots; Trevor Murphy.

For proofreading, many grateful thanks to Katrine Robinson, Maggie Ewen and Jo Donachie.

I would like to thank Jennifer Guest for creating a beautiful illustration for the front cover.

To Verity and Leonie, you are my own personal inspiration of what you can do with passion and talent and commitment. And love and thanks to Colin, who has been by my side and seen all the ups and downs of this immense project, and has variously encouraged, advised, sympathised and laughed. He has also dealt with images, researched and drawn maps, and brought visual clarity and consistency to the contents. This book wouldn't exist without you.

WITH GRATITUDE

Grateful thanks to the following RSGS members and supporters who have helped to make the publication of this book possible:

Eileen Abbess
Ian Adams
Sara Alexander
Doug Allan
Cath Alldred
Tim Ambrose
Fiona Anderson
John S Anderson
Nigel and Linda Armstrong
Matthias & Stefanie Berghaus
John D Birrell
Lynnette Borradaile
John Briggs
Colin Brown
David Brown
Jim Bruce
P & F Bryers
Duke of Buccleuch
Erica Caldwell
Nicola and Sam Chakraverty
Hugh Cheape
Aidan Christie
Dave Connor
Dr Alison Cook
Helen Cooper
Isabel Crawford
Roger Crofts
Nicola Crosbie
Simon Cuthbert
Andrew Dalgleish
Anne Daniel
Dorothy Degenhardt
Alastair J Douglas
Calum T R Edwards
Kevin J Edwards
Maggie Ewen
Denise Fallas
Fiona Findlay
David Flint

Ruth Fraser
David Geddes
Catherine Gemmell-Simpson
A P Gooch
Andrew Gordon
Alison Grassick
Ronald H Guild
Hamish Hamill
J W A Hamilton
Dr Ron Harkess
Alister Hendrie
Jasmine Hope
Robbie Horne
Dick Jennings
James Kennedy
John Kerr
Roger Kirby
Jamie Korner
John and Lois Lewington
Barbara Aldona Linder
Katharine M E Liston
Margaret Lobban
J M Loughridge
Helen and Burnett Lunan
Robert MacAndrew
Barbara Mackay
Morag MacKenzie
Iain Mackintosh
Elizabeth Mailer
Michael Mason
John McCutcheon
Martin McKay
Erica McKellar
Alan McVey
David McVey
Mary Murray
Barbara Noonan
Frank Norris
Francoise Orr
Dr Ruth Payne
Barbara Rasmusen
Dr Fiona Reid
Graham Ritchie
Alison Robertson
Allan Robertson
Luke and Hazel Robertson
Lynne Robertson
Jamie, Fergus & Euan Robinson
William D Roe
David Rowlands
Judith Roxburgh
Phyllis Runciman
Rosie Rutherford
Shiona Scott
Lynne Shiach
Donald Smith
Ian Snodgrass
The Sorrell Family
Sue Stockdale
Jeffrey Stone
Betty Stowe
Munro & Harris Strachan
David and Britta Sugden
Carole Tailford
Bob Thomson
William Valentine
Sophie Walker
Roger R Watts
Evelyne Wells
Miss S Whitley
Mr W G Whyte
Margaret Wilkes
Irene Williamson
Alison Wilson
Mary F Wright

PHOTOGRAPH CREDITS

For the images in chapter headings

Page 7: Henry Maull
Page 17: J Thompson, London, from *First on the Antarctic Continent: Being an Account of the British Antarctic Expedition 1898-1900*, by C E Borchgrevink
Page 27: RSGS Collections
Page 39: Hugh Simpson
Page 47: Eddie Fitzpatrick
Page 57: Scientific Exploration Society
Page 65: Benedict Allen
Page 77: Dougal Gray
Page 87: Rasmus Weng Karlsen
Page 95: Woodlore Ltd
Page 105: Royal Geographical Society (with IBG)
Page 115: Monica Jackson, courtesy of Simon Jackson
Page 125: Chris Bonington Picture Library
Page 135: Nick Yates
Page 143: Jessica Houlding
Page 155: Special Collections & Archives, Newcastle University
Page 165: Wellcome Images
Page 175: Rory Stewart
Page 185: John Smith
Page 195: Alice Morrison
Page 207: William Abbott Herdman
Page 217: Estate of Russell Westwood /National Portrait Gallery, London
Page 225: Doug Allan
Page 235: Michael Booth
Page 243: Mike Coots
Page 253: Office of UN Assistant Secretary General & Deputy Executive Director UNEP
Page 261: Solar Impulse
Page 271: Gary Knight
Page 281: Squiff Creative Media
Page 291: Gordon Buchanan
Page 303: Mike Robinson

Maps have been created by Colin Woolf using multiple layers from several sources with additional hand-drawn details. They are intended as a visual representation only and reflect original details mentioned in the text such as place-names which may have changed in recent times.

Every effort has been made to verify copyright holders of images reproduced. If any errors have been made please contact the publisher.

INDEX

Page numbers in **bold** refer to chapters